Climate Change for Football Fans

A Matter of Life and Death

Climate Change for Football Fans

A Matter of Life and Death

James Atkins

Illustrated by David Mostyn

UIT
CAMBRIDGE, ENGLAND

Published by
UIT Cambridge Ltd.
PO Box 145
Cambridge
CB4 1GQ
England
Tel: +44 1223 302 041
Web: www.uit.co.uk

ISBN 9781906860356

This is a book about climate change policy, which is one of the most boring topics in the world. So it also includes stuff about football, which is the most exciting topic in the world.

Contents

Part I

August to December

Climate change is as much about power stations
as football is about wearing shorts.

1. In the pub

A bet is made between two strangers with different interests.

No-one in Burnley had heard of Zaqatala until we put two past them the week before. And there was disagreement whether Azerbaijan was in Europe anyway. So instead of making the trip everyone piled into the pub to watch the second leg on telly.

It was a hot August evening at the Bridge, and Joe, Doris, Frank and the lads – Chopper, Baz, Spike and Beanie – were glued to the big screen, hands around pints.

At the table next to them an old geezer and a young bloke were talking earnestly. The older gentleman was Professor Igor Rowbotham. The younger chap talking to him – that was me.

Old Igor was in his eighties and he made me think of that Christmas carol. He spoke eight languages, owned seven passports, had degrees from six universities, had been through five wives in four continents, owned three cats, had fought in two wars, and had one interest: how to cut emissions of greenhouse gases.

He'd just arrived back in Burnley after forty years living in Patagonia. As he had no fixed abode, I'd agreed to put him up … for the time being. He'd been my dad's best mate in Burnley when they were kids before the war and they'd kept in touch over the years. His own dad was Burnley born and bred and caused a scandal when he married an exotic beauty from Ukraine.

Igor was having a bit of a grumble.

"Our failure in the last twenty years to take serious measures to cut greenhouse gas emissions is the suicide of the human race. We had plenty of notice. But we couldn't convince the man in the street. Most people on the planet don't and can't care.

"They say: 'Five degrees warmer? Fine by me. We won't have to fly to Majorca, we can sun ourselves in Margate.' A few hundred parts per million of carbon dioxide in the air? There's 50,000 parts per million of alcohol in beer and it has scarcely any effect at all. It just makes us pee a lot. So what is all the fuss about?

"This is as far as society has got since the 1980s when it was clear that global warming was a serious problem. Twenty years of reports, negotiations, political debates, films, public awareness, advertisements, hurricanes, drowning polar bears, disappearing glaciers, Africans, and women with Burberry headscarves fighting windmills. It has barely dented the psyche of the people in the richest countries of the Earth. Good people have been too good, and bad people too bad."

Suddenly he gripped my arm and looked towards Joe and the lads at the next table. "Listen to them," he whispered. "They're talking about it."

"It fucking gets up my nose," Chopper was saying. "They're always bloody going on about it. What do they want us to do? Live in a fucking cave?"

"It's just lefties and commies, isn't it, looking for summat to whine about…" said Beanie.

"Chantelle's always bollocking me about the recycling," added Spike.

"For a start if scientists don't agree, why the hell should I bother?" said Frank.

Joe interrupted the debate. "Can you all shut up now, they're back on again."

Then Beanie spoke up again. "If big bloody business put it up there, don't you think big bloody business should put it back again?"

"And the Indians," said Spike. "Who saw that slumdog film? Just look at them."

"It's the Arabs. It's all their fault," said Baz.

"Can you all shut your bloody gobs please and watch the match," shouted Joe.

"That's the thing the Americans didn't sign. Why bother, that's what I say," said Beanie.

"We don't do spray-on any more. Just them roller-ball things."

"We knew this family in Blackburn. They died of carbon dioxide poisoning. Lovely family they were."

"It's all sunspots anyway."

"For the last time can you all shut up – please! I am trying to watch a game of football," screamed Joe.

"No-one is going to tell me where to spend my holidays. Fucking nanny state," added Frank quickly.

"And the Indians. I reckon we should nuke 'em. No seriously."

Burnley scored again. There was joyful pandemonium.

When it was all over the Professor tut-tutted. "What a terrible thing football is."

"I know, 22 men in shorts rushing around after a little leather ball," I joked. "Come on, Professor, it's fun and it's better than them beating each other up and taking drugs."

"That may be. But it's vulgar, dull, and it provokes the worst behaviour possible in people. I've no interest in football whatsoever."

"Hey," came a voice from the next table. We looked up and saw that Joe there was alone with his pint. He looked at the Professor. "Did you just say that you have no interest in football?"

"I certainly did. It's vulgar and utterly tedious."

"On the day that the Clarets beat Zaqatala four nil away? Tosser."

"I am sorry, young man. You've lost me. Clarets? Zaqatala? Tosser?"

Joe sighed. "Right. The Clarets are Burnley – the best football team in the world. Zaqatala are a bunch of wankers from God knows where. Tosser? You know what a wanker is, right? Well, a tosser ... that's ... er ... like a wanker."

I made the introductions. "Joe, this is Professor Rowbotham. Igor Rowbotham. He's just ... er ... moved here from ... Patagonia. Professor Rowbotham, this is Joe. Joe Sugden. Joe and Doris also live on Juniper Close."

"If you don't like football, what do you like?" Joe asked.

"Studying how to reduce anthropogenic emissions of greenhouse gases," replied Igor.

"Bloody hell! I've heard of Jan Vennegor of Hesselink and

Edson Arentes do Nascimento, but that takes the biscuit."

"Well, you've heard of climate change?" he asked.

"Course I bloody have. That's what they were all going on about when I was trying to watch the match. But it's a waste of time, isn't it? You'd be way better off spending your time watching Burnley."

Far from being offended, Igor fetched a fresh round for everyone.

"Joe," he said when he returned. "Your passion is the fate of Burnley football club. My passion is the fate of the earth. We are both passionate and committed men. I expect that you have encyclopaedic knowledge of Burnley?"

"Try me," said Joe.

"I wouldn't know what to ask."

"What position did Burnley finish in 1965?" I asked.

Joe thought for a moment. "Twelfth."

"Who was top scorer in … 1981?" asked the Professor.

"Billy Hamilton, of course.

"Extraordinary," said Igor, turning to me. "He seems to know everything about them."

"Sure I do," said Joe. "You would if you'd been brought up on it like me. Course, it's too late for you now at your age."

"Too late for me? Cheeky whippersnapper," said the Prof. "I've much better things to do than waste my time learning about football. It'd be like you taking an interest in climate change." He chuckled.

"Why not?" Joe bridled. He must have been on to his fourth pint. "Why shouldn't I take an interest in it? Not against the law is it?"

The bet was made on pint five. "I bet you," said Joe with a slur, "I bet you I can learn more about your climate crap in a season than you can about football. Thousand quid." It was a moment of bravado.

"Don't underestimate me, young man," said Igor. "I might be old-fashioned and grumpy and not always hold in my farts, but there's a brain in here," he said tapping his head. "I'll take on your challenge. Our friend here can be the judge. I'll jolly well get a season ticket and come with you to every Burnley game. And … meanwhile you'll work with me on the question of what politicians can do to reduce anthropogenic emissions of greenhouse gas."

"Except if you talk like that, we're not going to get nowhere," said Joe.

"In plain English. Deal?"

They raised their glasses. That was how it began.

2. Getting started

Every little helps, but only a little.

Igor had this way of barging into people's lives and making himself feel at home. I'd been working at a local building society, with a Ford Focus, a long-term bird and a season ticket to Burnley FC. Then along came the recession and I lost my job, a van in a hurry trashed my car and some arsehole on 50k who supported Chelsea nicked my bird. Well, at least no-one took my season ticket. Then the Professor turned up on my doorstep. And now he was doing this barging in again with Joe.

The next Sunday we all met up at the Bridge. It was the charity shield match between Manchester United, the champions of the Premier League, and the FA Cup winners, Hull City.

Joe brought his wife Doris along, and their two teenage children Darren and Kelly. Everyone was wearing a Burnley shirt.

Igor made an earnest start to the game, jotting things down in his note book and squinting up at the big screen telly. "I can see what's happening, but it's like an orchestra playing with no music coming out. I see no beauty there, no reason. Just motion."

Joe frowned. "But look at that move, there … Look how he pulls away the left back, making space for Giggs …"

"I see movements, my friend. I see no system. Just figures drifting across the screen. And the ball." He paused. "Interesting … the ball … does the writing on the ball signify something? There's a pattern of lines on it. What does that mean?"

"Nothing! It's not important. It's just something the manufacturer put on it. Now look there. See how Nani's holding the ball … look how he turns into space …"

"But they've suddenly stopped running around!"

"It's bloody half time, they're going off for a rest. They have oranges and a cup of tea at half time," explained Joe.

"When do they start again?" asked Igor.

"In fifteen minutes."

Igor looked at his watch.

Joe took a swig of beer. "Looks like it's your turn, now, Prof. Let's hear what you have to say."

"Thank you, my friend." He coughed and looked round at the family waiting expectantly. "Now, I'm afraid some people find my topic rather dull."

"Mm … sounds fun," said Darren poe-faced. Kelly giggled.

"Don't be rude, Darren," said Doris.

"But they simply don't dig deep enough. Look, I studied cosmology for thirty years. It never ceased to fascinate me … discovering the secrets of the universe through numbers."

"Cosmology?" said Darren. "You mean … spacemen and stuff?"

"I mean the very first moments of the existence of the Universe."

"Right …" began Joe. "Fat lot of use in that, Professor. What's like the point of it?"

"It has no point."

"So why did you do it?"

"I enjoyed it and got paid for it."

"Nice one," said Joe. "Imagine getting paid for going to Turf Moor."

Doris said: "So why are you worrying about all them greenhouse gases then? I mean, if you've seen all them stars and stuff, doesn't everything else all seem a bit …"

"You see, I had been looking at the sky all my life." He shrugged. "Must have had a stiff neck."

"Darren!" said Doris sharply. "I'm sorry, Professor."

"No, that was the problem," said Igor. He stood up and rubbed his neck. "I had two weeks in a sanatorium with my vertebra. And as I lay in bed I looked out of the window, and I slowly became aware of the earth around me. I had completely lost touch with the planet I was sitting on. Everything was numbers and theories, computer screens and stars and telescopes. Things a very long way away and long, long ago. It exercised the mind, but … and it ensured that the minds of my students were exercised. But in that hospital ward, springtime, windows open … I started to hear bird song again, to notice the rain, to see colours around me … Sorry, I'm distracted." He paused.

Cheers went round the pub as the players took to the field again. "Ignore them," laughed Joe. "It's some twats who support Man U. They shouldn't be allowed in here. We'll have to carry on later," said Joe, "they've started again."

"You have penalties," explained Joe, fifty minutes later, "at the end of some games. If it's a draw, and you need a winner, then you have penalties at the end to decide it."

Hull's penalty kicking proved more determined than Manchester's.

"And how many points will Hull get for this win?" asked Igor earnestly.

Everyone laughed. Doris took pity. "No, this wasn't a league game. This is just a one-off match. It's not part of the Premiership."

"I see. Still, I don't understand why Hull won. Aren't Manchester a more successful team than Hull? How come they scored fewer penalties?"

"Penalties are pretty random," said Darren. "It's not about how good you are … penalties are about your psychology. They're under pressure. They concentrate too much. You start getting demons. And you make a hash of it."

"Ah, my friend! Psychology. Greenhouse gas emissions are also pure psychology. It's all about the mind of the human being."

"You what?" said Joe, taking an interest. "Climate change is all about psychology? I though it was about aeroplanes and recycling. You know, windmills and that kind of bollocks."

"On the surface, yes. Perhaps that's the dull bit. Technology, industry, taxation and finance. Very dull; we'll skip over that. Once we dig deeper, it's about psychology, evolution, neurology, biology, and … well, sex."

"Sex?" cried Darren, alert. A hundred faces turned towards him.

"You see," said Doris. "I said it'd be interesting, Joe."

"Another beer, Prof?" asked Joe hurriedly, picking up the empty glasses.

The Professor sat back with his fresh pint.

Joe said: "Er … I don't want to waste your time. I mean, we do know about this stuff. It's on TV all the time. Doris is always going on about her carbon footprint. She's always reusing plastic bags and turning the lights off."

Igor shook his head. He wiped the foam from his white moustache. "A complete waste of time."

"You what?"

Doris bridled: "Are you saying it's a waste of time to use plastic bags and turn your lights off? How can you say that?"

"I'm sorry, my dear. I didn't mean to upset you. But think of plastic bags. Manufacturing a plastic bag causes emissions of 20 to 30 grammes of CO_2." He took a deep sigh. "How many times do you go to Tesco a year?"

"Once a week."

"Let's say you use 5 bags a time?"

"I dunno. Maybe."

"So that's 250 bags a year?" said Joe.

"Not bad, Dad. Mr Numbers from Planet Maths."

"Watch it, Darren, you'll be paying for the next round,"

warned Joe.

"250 bags a year. Meaning 6 kg of carbon dioxide," said the Professor with a smile.

"Well it's better than nothing," sniffed Doris.

"Do you know how much emissions come from your household, each year?" Igor continued.

"Not quite sure," said Joe. He'd got distracted by the highlights. "I think I heard, like … could it be a ton of … you know?"

"A ton of?" Igor raised his eyebrows.

"Well, carbon … carbon dioxide … er … or is that carbon monoxide?" Joe hazarded.

"Carbon dioxide, my friend." The Professor opened another pack of pork scratchings, offering them round. "It could be 10 tons … in a year. For a person. 10, 15 … depending on how you calculate it. Per head."

"10 tons? But that like weighs …"

"Yeah. It weighs 10 tons, Dad," laughed Darren.

"Yeah, but, I mean …like … I weigh 90 kg … How's that compare?"

"It's over 100 times your weight," replied the Professor.

"So in one year, like, I cause a 100 times my own weight in carbon dioxide?"

"Probably."

"Wow. And millions and millions of people are doing that all day?"

"Billions, my friend."

"Jesus. You don't think of it like that, do you? I mean if it weighs 10 tons, how come it doesn't just come back down again? I didn't think gases weighed anything."

"I think," said Igor, "that we should start at the beginning" Then he added: "There might be times when what I have to say isn't so exciting. Just like sometimes I'll have to sit through games without goals. But believe me. Like football, it's a matter of life and death."

3. People not machines

At the root of emissions are the motivations of people, not machines.

"This may take some time," warned the Professor. "We have much to discuss. And your knowledge of the topic is similar to my knowledge of Burnley FC. Close to zero."

We'd all been invited to dinner by the Professor to the Bay Horse in Fence, probably the best restaurant in the Pendle area. Joe was tucking into his T-bone steak and Doris was struggling with prawns. Igor wolfed down his salad, folded up his napkin and got straight to business.

"Let us agree that climate change is a serious problem. And let's agree that it's in part caused by man-made emissions of greenhouse gases."

"Whatever you prefer, Prof."

"And I'll assume that football is a worthwhile activity, and one enjoyed – genuinely enjoyed – by millions of players and spectators alike."

"It's the best followed sport on the planet," said Joe. "Did you know that there are 192 million Manchester United fans in Asia? And they don't even know where Manchester is."

"The Chinese are heavily involved in this one, too," Igor said. "Although luckily the Chinese have a far better understanding of this problem than they do of the geography of Northern England.

"Let us start with reasons. To be able to do something about emissions, we have to know why they happen…"

"Fair enough."

"One way of looking at the causes of man-made emissions is how the IPCC does it."

"IPCC? What's that? Some cricket team?" asked Joe.

"The IPCC is the Intergovernmental Panel on Climate Change. They see emissions as coming from different machines: power stations, boilers, cars, buses, factories; and from different places: waste tips, farms, forests, fields."

"Right," said Joe. "It just looks at what happens to make the emissions. It's like saying that football is twenty two men running about in shorts – "

"Exactly my friend," said Igor. "Another way of looking at emissions is to look at what causes them. Why did those power stations need to generate so many gigawatt hours of power? Why do Americans consume 8 million barrels of oil everyday in their cars and trucks? You can say that all this burning of things and processing of chemicals and tilling of the land happens because of the things we do: making and eating food, keeping our houses warm, having baths, getting to work, curing our diseases, entertaining ourselves, keeping the army and police and so forth. This way of looking at emissions focuses on the end-user. It's us."

"Right," said Joe. "This is a bit like … why are we running around after the ball … to get it into the opponent's half … to get it away from our penalty area … to get it into their goal…"

The Professor nodded. "And you can go a step further. One direction is to ask what are the underlying needs which the things we do satisfy? Sometimes they serve important biological or physical functions – like procuring food and eating, keeping clean, keeping warm, or keeping safe. Sometimes they serve psychological functions like giving us status, security, recognition, love, making us feel good or happy or excited or safe, or giving us a feeling of novelty.

"Another question to ask is … why do we do things this way and not that? What are the underlying motivations to use the car and not just stay at home or make the effort to take the train?" said Igor.

"You mean why play football at all?" asked Joe.

"Indeed. You could go further and look at our underlying instincts … the need for competition, the need for exercise, the desire to fight and win … and the projection of those tribal instincts into a game … since we don't need to use the instincts any more just to survive."

Joe nodded uncertainly. "I kind of get it, Prof."

"Good, my friend. This is a simple illustration about how emissions are ultimately caused by our efforts to satisfy biological and psychological needs. So, how do we tackle emissions?"

"Dunno," said Joe. "Sit in the pub and do nowt all day?"

"Maybe. So far policies are almost all about improving our machines. But we're dealing with one of the hardest, biggest and most urgent problems in the world. It would be sensible to be open-minded in our search for things to do about it. Why don't we look at the real causes of emissions?

"Remember that we are talking about man-made emissions of greenhouse gases. These emissions aren't made by baboons or earwigs."

"That's right," said Doris, finished with her prawns, the heads laid neatly around the edge of her plate. "It's seventy-odd thousand Burnley supporters watching the match in Zaqatala on TV."

"Exactly, my dear, " said Igor. "Behind every kilogramme of carbon dioxide emitted there's a human action, a human decision, a human motive. This is very important for us to bear in mind."

"You mean," blurted out Joe, "it's about people, not technology and machines. It's like if Burnley want to stay in the Premiership, it's not just about tactics and stuff, it's about belief and passion … and motivation … what goes on in your heart."

"Indeed, my friend. Policies for people. Policies of the heart and mind. More wine?"

4.   Dimensions

Some numbers to give context.

The opening game of the Premier League season was against Newcastle. At Igor's insistence we'd taken the train all the way only to lose 1-0. It was a nervous game with little action and not the best way to introduce the great Professor to the thrills of football.

On the way back from the north east, Joe started with the basics. It made sense to begin with simple dimensions because the Professor was so good at numbers. The match is ninety minutes long. Two halves of forty-five. Two teams with ten outfield players and one goalkeeper. You can have seven subs. The goal is eight yards by eight feet. The pitch is between 50 and 100 yards wide and 100 and 150 yards long. The capacity of Turf Moor is 22,000. "You have to learn this stuff, Prof. It won't help you love Burnley. But you still need it."

Igor dutifully noted all these statistics at the back of his book. He now flicked through to the front.

"I'm afraid the world of emissions is a horrible jumble of numbers, too. Grams of this, kilograms of that, tons and megatons, parts per million, percentages and probabilities, CO2 and CH4 and SF6 and HFC23 …"

"Bloody hell, I'm not supposed to learn all that, am I?" gaped Joe. "I'm shite at maths."

"No, no," said Igor. "It's very simple. The UK is responsible for just over 900 million tons of greenhouse gas emissions. There's 600 million tons which are created within the UK and there's another 300 million tons of emissions caused by all the stuff we import from other countries. So if there's 60 million people and 365 days a year, it comes to 42 kg per person per day."

"Hey wasn't that the number in that Hitchhiker's Guide thing? 42?" said Darren. "Bloody scary, that."

"Now, usually people break down these emissions by saying that so many come from power stations, so many from cars, so many from agriculture and so on. But remember that every ton of emissions comes from something which people do. One way of looking at them is to break them down into three categories."

"Go on," said Joe.

"First, things which we do every day. Which cause lots of small emissions day in day out."

"Like … having a bath?" said Doris.

"Exactly."

"Or having a cuppa. And driving to work," said Joe.

"Perfect. Then next group is special things we do from time to time ..."

"You mean like ... going out to dinner with the Mrs?"

"I'm thinking more of things which have a big impact on our emissions but we do them rarely. Things like going on holiday. Or more like buying a new fridge and getting a new car. Making these things causes a big guff of emissions. And then the catch is that once you've bought your car or your freezer, it just goes on causing emissions for years after that. Most unfortunate."

"Hmm ... 'most unfortunate'," repeated Darren. "That's 'double whammy' to you, Dad."

"Those two groups were things we can do something about. Then there are things which we can't do much about. Things that are done for us, which we benefit from, but we have no influence over."

"What do you mean?" asked Joe.

"Emissions from fighter planes, from government buildings and the tax office, from the police, from county council offices, from maintaining the country's infrastructure–" Just then the carriage lurched and the train came to a stop.

We were in the middle of the countryside somewhere on the Yorkshire moors. We looked out across sweeping, treeless hillsides cast with the evening light. A wind had blown up, buffeting the rooks which were making their way home to roost.

"If I'd known that you'd make us take the train everywhere, I'd have had second thoughts about this lark."

"Second thoughts? Just look outside! Look at the beauty of it all! The wild landscape... The purple of the hillsides. The bracken. The wandering sheep. Treasure it!" Igor said.

"Bloody wandering sheep. It's Yorkshire, isn't it? Anyway I'm gasping. Doris, have you owt left in the thermos?"

"Sorry love, thermos is empty. But..." said Doris.

The Professor stood up and hauled his bag down from the shelf. "Perhaps these would be more suitable." He pulled out six bottles of Newcastle Brown.

"At least we took something away from Newcastle," mumbled Joe.

"Now, where were we," said Igor, sipping from his bottle.

"You had your day-to-day stuff," said Darren, looking up from his Nintendo. "And then your one-offs. And then the stuff done in the country that you can't do anything about. And – "

"Haven't you something to go with this, love?" asked Joe. Doris magically produced a Victoria sponge.

"Perfect," said Igor, delighted, as Doris laid the cake out on the small table in front of them and moved to cut it. "No, no," Igor said, taking the knife out of her hand.

"This cake represents 900 million tons of carbon dioxide and other greenhouse gases. All the emissions produced in the United Kingdom and all the emissions from things made elsewhere which we import and consume here. All the emissions this country is responsible for. And ignoring emissions from things which we make here and export.

"Now these," he said, cutting a huge slice over half the cake, "these are all our everyday things – cooking, washing, heating, eating … and so forth.

"Then this piece … this is our emissions from specials – leisure and entertainment, from buying new cars and gadgets …"

"This third piece," he said, prodding the final slice in the cake, "is the emissions from defending and administering the country."

"So most of them are things we're doing in our everyday life," Joe remarked. "Just there are lots of us doing them."

"Sixty million people doing these things everyday," said Darren.

Joe helped himself to a hunk of everyday emissions. He offered me the slice of specials. Leaving Darren and Kelly to fight over the administration of the country. Doris wasn't eating cake and Igor had overdone it on veggie burgers at St James' Park.

"It's a good job that everyday emissions are so high," said Joe through a mouthful of jam and sponge.

But he was soon hungry again. "You can have some Smarties, Dad," suggested Kelly.

"Perfect," Igor said, pouncing on the coloured chocolate buttons. He poured a pile of Smarties onto the cloth on the table and began count them and sort them into colours. Then laid them out on the table like this:

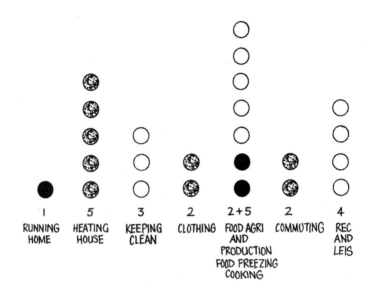

"Here are 24 Smarties – 24 kg of carbon dioxide per person per day from the things which we do every day. It's a rough average."

He pointed to the short column on the left. "The 1 kg from running the home – lighting, vacuuming and cleaning. Next to it are another 5 kg of carbon dioxide from heating the house and other buildings. That's emissions from the boiler burning gas. That is per person."

The third column is about keeping clean – 3 from washing and showering – all that hot water, running the shower motor and so forth. And the fourth is from clothing. Some is running the washing machine, ironing, drying and the rest is a portion of all the emissions from making the clothes you wear."

"Emissions from clothes?" said Darren. "What you on about?"

"Making fertiliser for the cotton fields. Tractors at the cotton farm. Transporting the cotton. Processing the cotton and the fabric. So if you add up 10 t-shirts, 5 shirts, 3 pairs of jeans, a few hoodies, a few pairs of trainers, your boxers, your socks ..."

"The Burnley kit," said Joe.

"A couple of baseball caps..." added Kelly.

"Scarf," said Doris.

"Hat."

"Gloves."

"You get about 600 kg of emissions from making all that... and they last about two to three years. So that's just over half a kilo per day for you, Darren. Luckily you're under the average. Think of people with eight or nine suits, twenty ties, several sets of leisure wear, a dozen pairs of brogues, spats, a trench coat, a formal overcoat, a mackintosh, three umbrellas, a summer dressing gown and a winter one ..."

"And all them women with a dozen summer dresses," sighed Doris. "You see, we haven't got anything like that," she added, glancing at Joe. "I'm very careful with what we spend on clothes." Joe raised his eyebrows.

The Professor carried on: "The biggest column is emissions from food. 2 kg is from cooking – the oven, the kettle, the fridge and freezer, the dishwasher and so on, not to forget the weekly run to the supermarket ... 5 more kg are from producing food – transport, factories which process food, tractors and combines at the farm, making packaging and so on. Everything – your pizza and pasta and butter, bread, milk, jam, Marmite, Rice Krispies and All-Bran Yoghurt Bites, balti takeaways, winter tomatoes from the hot house, strawberries flown in from South Africa ... That includes emissions from agriculture – which are all part of getting food onto this table. About 2 kilos per person per day."

"From agriculture?"

"Oh yes. Methane coming out of cows and other farm animals..." Kelly giggled again.

"At both ends, right?" asked Darren.

"If you will," replied Igor. "And gases which come out of the ground when we farm it, especially when we farm it intensively: more methane, nitrous oxide, and carbon dioxide. In fact this number is a bit light. It ignores the emissions from farms overseas where they burn down the forests to make fields for cattle to graze on."

"Do you think ..." began Joe.

"I can do without this one," said the Professor and shoved a single pink one over to Joe. Joe gobbled it up. "Dad! That's not fair! they're my favourites, the pink ones!" protested Kelly.

"What's that pink column, then, with them two in?" asked Darren.

"That's 2 kg of emissions from commuting. 8 or 9 kg for the household each day."

"And this last column is for recreation and leisure." Igor pointed to the alternative purple and blue Smaries. "Approximately 4 kg."

"What's in all that?" asked Darren.

"All the electricity used by your television, your computers, your hifi, radio, and play station. All the energy to make books and newspapers. A bit more than 2 for all the travelling around we do for leisure – driving to matches, to the cinema, to dancing lessons. It's your share of all the emissions from leisure activities going on around the country – cinemas, theatres, leisure centres, sports clubs ... and so on. And your Newcastle Brown, of course."

"For heaven's sake, don't keep going on about Newcastle, will you," said Joe.

"So that all makes 29 kilograms per person per day. Just for the everyday activities."

"Is that it, then?" asked Darren. "Can we have the rest of them?"

"Certainly not. That was just the daily routine things. Then we have specials."

A bit more quickly this time, he shovelled together a pile of 12 more Smarties.

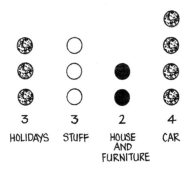

"The 3 yellow ones are emissions from holidays, especially from flights. It's the trip you make to Spain every year … something over 300 kg for one round-trip. And Joe's trip sea-bass fishing in Florida with his brother. That's well over a 1,000 kg, a ton … So altogether it comes to another Smartie per person per day.

"The 3 red ones. All your household goods. Dishwasher, washing machine, cooker, expresso machine, vac, fridge, freezer, moulinex, tumble drier, TVs…"

"Gameboy?" asked Darren.

"Of course," said Kelly. "Iphone, mobiles, iPod, laptop, computer… Like all the stuff you need for basic living, like."

"Exactly, my dear." Igor smiled. "All this stuff has to be made somewhere. Drilling the oil in Saudi Arabia and mining ores in Australia, processing them into metals and plastics, working them into parts, assembling them, transporting them criss-cross from factory to factory down the Yellow River, up the Yangtse, down the Changxing-Shanghai canal … then across the Pacific … down the Mississippi … over the Atlantic … and finally up the M6 and down Juniper Close."

"Fu–," began Darren. Doris slapped him.

"And all that's 3 Smarties a day per person. Roughly. Of course, it depends on how much stuff you have. It's a rough average."

"And the 2 pink ones?"

"That's all the emissions from building your house and the

furniture in it. Divided up over the life of the house. Then the purple ones. That's for manufacturing the cars. About 3 kg a day from yours, and a bit less from Doris's."

"Bloody hell," said Joe wiping his brow. "There's no end to this."

"We're almost there," the Professor said. He set down the remaining 6 Smarties.

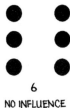

6

NO INFLUENCE

"These are the emissions which you don't have influence over. The government's activities, the defence of the nation, healthcare and education services, all the communications going on around the country which you're connected to."

"Well at least I can eat those ones up," sighed Joe.

"So you've got the picture? That's 42 kg of emissions on average for each person. About 24 kg which are just things you consume every day, another 12 kg of specials – treats, holidays, and the big things you buy ... and another 6 kg from the running of the country. It's more complicated than that really, because of imports and exports, and uncertainties, and ways you ... Well, it's a rule of thumb."

"42 kg" said Joe. "It still doesn't mean much."

"It's about what Kelly weighs."

"So each day, she emits her own weight in greenhouse gases?" laughed Darren.

"Indeed."

"Blimey," said Joe. "You wonder where it all goes."

"Duh! Into the sky, dad. That's the whole problem," said Kelly.

"Dad, you can teach the Professor about formations now,

with these," said Darren, eating the commuting emissions. "4-4-2, 4-3-3 ... and all that. You only need eleven, so we can – "

"Professor," asked Doris, as Joe scooped up the rest of the Smarties and stuffed them in his mouth. "What should the emissions be, like? I mean if we wanted to be safe? Like, really safe."

3

TARGET

The Professor picked up some Smarties which Joe had left.

"Three?" exclaimed Darren. "We've 42 and have to get down to 3 kg a day? But that's like more than 90%."

"Yes, one ton a year. It's about 3 kg a day per person. That's all the planet can manage. Remember, there will soon be over seven billion people."

"Bloody hell," said Joe. "From 42 down to 3. How are they going to do that?"

"That's a very good question, my friend. No-one really has a clue."

5. Finding the reasons

The Professor interrogates Joe to find the causes of emissions.

Igor said that he'd need to spend a lot of time with Joe collecting information. We attended the midweek match in Birmingam – beaten 1-0 and a dozen missed chances. In truth we defended well and there was no shortage of attacking. But when it came to the crunch, Eagles scuffed, Fletcher miskicked, Patterson angled it wide and Eagles's header flew just over the bar. The ball just wouldn't go in.

On the coach back up the M6 Joe grumbled to the Professor that the Clarets weren't fighting. Next May seemed a long way away. "They think they've got time. But every game counts. Three points may not seem much now, but by the end of the season it's the difference between life and death."

"Those misses are all in the mind," said Igor. "It's not about effort or ability. It's what's going on in their heads. In fact," Igor continued after a moment. "I wanted to ask you some questions."

"What about?" asked Joe.

"About heating. About eating. About travel and leisure. About how you live. About the things you buy. About why you do one thing rather than another. Let's start at the beginning."

The Professor grilled them the whole way from Birmingham to Burnley. He asked them about the temperature in the bungalow in the summer and winter, about how many jumpers they wear inside, about the shapes of their radiators and the size of their boiler, about thermal underwear and bed-socks, about how often they have baths and showers and how long they take, about how often Joe puts on a clean shirt.

"He stinks after a day at work," explained Doris.

"It's a man thing," said Kelly.

"The washing machine. I saw that you had it on a 60 degrees wash," said Igor.

"Really?" said Doris. "I never know what I have it on. Can't do with always fiddling about with the knobs. Anyway 40 degrees doesn't clean properly."

Doris looked around at Joe for support. But Joe was distracted by something on the coach TV.

"Joseph," called the Professor. "My friend! Please pay attention! We will get to the television in good time."

The Prof went on to ask about eating and cooking. Joe and Doris answered the best they could and Igor jotted everything down in his note book.

"This filling the kettle full – why not just boil as much water as you need?"

"Never thought about it," shrugged Joe.

On the coffee habit: "We don't like apple juice. Makes your teeth go funny. You need something to get you going in the morning."

"And when you've finished a little job," said Doris, "you need a reward."

On meat eating: "Vegetarian? Lentils and shite? You want me to wear daft hats, too? Anyway, I need the protein to make my brain work."

Doris said: "I haven't always got time for cooking, have I? What with the kids and my job and the house work. Microwave food's easy, isn't it?"

Then they got on to free-time and leisure. He asked about morning telly and whether the telly was on in the kitchen during breakfast.

"Of course," said Joe, surprised. "How could you watch GMTV in the lounge if you're having breakfast in the kitchen?"

"Silly me," said Igor. "And do you switch it off again after it's finished?"

"Dunno. I think it's just on all day, isn't it, Doris?"

"Well, I wouldn't say all day. But it's nice to have a bit of company when you're at home."

"So you have a television in the kitchen, one in the lounge …"

"Yes that's right," said Doris. "And one in our bedroom, and one in the kids' rooms. And one in the spare room."

"One in the bathroom?"

"No, not there," replied Doris.

"In the garden shed?"

"No, not there either" said Doris looking concerned.

"In the garage?" asked Darren.

"They're pulling your bloody leg, love," laughed Joe.

"Now, do you always have the television on every evening, or was last night an exception because of the football?"

"Me, I just watch the football. And Road Wars. That's it," said Joe.

"Oh really, Joe Sugden?" said Doris. "What about Strictly? And Top Gear?"

"Right, sometimes," said Joe sheepishly. "Well look at Doris. She watches Corrie, and Eastenders, and X-Factor, and Britain's got Talent, and Saturday Kitchen."

"And Antiques Roadshow, Jonathan Ross, and Question of Sport added Darren.

"Well I know it's all rubbish," said Doris. "We could do without it completely if we really wanted."

"Except for the football," said Joe.

"And the nature programmes. They're beautiful." said Doris with a pert smile.

"Well, I never watch telly," said Kelly.

"Yeah right," said Darren. "Generation Facebook."

They moved from telly to other leisure activities. Monday was Eastenders, Tuesday five-aside football in Brierfield, Wednesday was ladies night at Hollywood Park for a film or a Chinese in Accrington, Thursday was shopping at the Trafford Centre, and Friday was pub night.

"At the weekends we might drive up to the Lakes for a toby," said Doris.

"Or just go to Boundary Mill and see what's going on," added Joe.

"What's going on? At a shopping centre?!!" he asked.

Then Igor asked what they'd do if they had all the money they wanted. Doris went for a week in Corfu, like the Millers did.

"I'd get a new Golf. GTi 2.0 TSI. Actually, I might get an Audi A6 like the Shaws at number 26. Posh buggers, but they've got a nice car," said Joe.

They'd buy Uncle Frank a golfing holiday in Spain, too. And get one of those American fridges and the Kenwood Titanium Chef food mixer they saw on Masterchef. And a pool. And a Panasonic flat screen TV they said would be in the Comet sale. Darren would get the iMac and Kelly would do extra dancing classes in Burnley. And they'd spend all Sunday at Boundary Mill.

They talked about transport. Igor asked about how Joe gets to work and Joe laughed and said by car of course. Why Joe didn't take the number 25 bus which would take him from the Bus Station in Burnley to the Bradford and Bingley in Colne? From there it was just a few minutes walk up to Craddock Road. Under an hour door to door. "Or about forty minutes on your bike," Igor added.

"On my bike? What do you think I am? A bloody student? This is Burnley, not bloody Oxford and Cambridge."

The Professor had one more question.

"Go on, mate. They're not showing our highlights anyway."

"The jet-ski. What is this about?"

"The jet-ski? Ah, come on. You need a bit of excitement in your life, don't you? What do you want us to do, go and become nuns?"

"You couldn't be a nun, Dad," said Kelly. "You'd have to be a monk."

As we walked home from the bus station, Joe asked Igor if there was going to be much more of this inquisition stuff. "It does my head in."

"We'll meet again before the game against Everton on Saturday," said Igor.

"Your first home game," said Doris. "Are you excited?"

"Excited?" said the Professor. "No more excited that you'd be, my dear, studying the Climate Change Act of 2008. But I will certainly bring the scarf you knitted, and will wave it when told to."

The Professor made a great effort on Saturday and waved the scarf from time to time. It must have worked, because we played a knife-edge nil-nil draw and won our first point of the season.

6.   In the mind

The Professor says that many emissions from our everyday activities are avoidable because they are just caused by things going on in our mind.

At the end of August we had the first leg of the play-off round of the Europa League. Cerno More Varna were making the trip from the Black Sea to Turf Moor. I was round at Joe's helping him with his pools and the Professor met us there before we went to the match.

He arrived breathless – banging at the door, a bottle of claret in each hand.

"It's all in our minds," he announced as he threw off his coat, missing the coat-peg by a foot. "All the things we do which use all this energy – it's all down to habits and things going on in our heads. We can get rid of most of it just by sorting out what goes on between the ears." He jabbed his forehead excitedly.

"Hold on Prof," said Joe patiently. "Get yourself sat down first."

I fetched some glasses, and poured the purple liquor. After a couple of gulps the Professor calmed down. Perched on a sofa, he began to explain.

"Look at eating meat. Harmless and traditional thing. We all enjoy it. You can't beat roast beef and Yorkshire pudding."

"It's the only thing from Yorkshire we ever have in this house," said Doris.

"So what's wrong with eating meat?" asked Joe.

"I don't want to be a killjoy…"

"But…" continued Darren.

"Exactly. So if we have to cut emissions by 80%, meat eating will have to become something just for special occasions."

"Special occasions?"

"Once a week at the most," confirmed the Professor.

Joe protested. "But I can't stop eating meat! It's not a proper meal if it doesn't have meat and two veg. I'm not some poof that does lentils and shite. A meal's not right if it doesn't have a bit of meat with it."

"So what about people who don't eat meat? Do you think that they don't enjoy their food?" Igor asked.

"How could they?" asked Joe.

"Of course they do," said Doris. "They've probably just got used to not eating it, that's all. People eat what they're used to, don't they? Look at the things foreigners eat."

"You see," said Igor. "It's just a question of habit, what we're used to. If we're used to eating meat every day, we feel it's right to eat it every day. If we're used to eating it just once a week, say, a good Sunday roast with that certain regional pudding, roast potatoes, carrots and peas, gravy ... then that's what we'd feel is right to eat."

"You must be right. It's the only way to explain why folk eat weird stuff," admitted Joe.

"And why people who only eat a little meat are still perfectly healthy. They don't actually need it," said Igor.

"So, you mean, it's just a question of what we're used to? Nothing more complicated than that?" asked Doris.

"Yes. Pure habit," replied Igor. "Here's another one. Heating. It's one of the biggest causes of emissions. And it's not about things people do in China or a power station miles away. It's here at home. It's that little box on the wall."

"The boiler?" asked Joe.

"No, the thermostat. Any idea what temperature you have it on in winter?"

"Haven't a clue," admitted Joe cheerfully.

The Professor got up and walked over to the thermostat on the wall. He peered at the small LCD display and pressed some buttons. "23 degrees in the winter," he pronounced.

"Come on then, let's hear what it should be," laughed Darren.

"How does 15 sound?"

"15?" gasped Doris. "We'd catch our death."

"No we wouldn't, mum. That's what he's saying. It's just what you're used to," said Kelly. "I bet there are loads of people who have their houses at 15 degrees and they never get colds and stuff."

"That's my point," said the Professor. "Having the house at 23 degrees – it's just what you happen to be used to. You can get used to practically anything. You don't need to have it at 23. You'd be fine at 15 degrees."

"If you wrap up warm," said Kelly.

"Or jump about a bit," said Darren.

"If you cut the heating from 23 degrees to 15 degrees, you'd cut your winter heating bill by half."

"You'd have to learn how to set the thermostat, though," said Joe. "And that means a bloody degree in computer programming."

"The thermostat's automated your heating, so it's become a habit to take no notice of it. Break that habit and set it often. Take it down half a degree each week. For six weeks. At least. No-one will notice. Especially," he said with a wink to Doris, "if you buy her a beautiful alpaca shawl to snuggle up in."

"Alpaca shawl? You must be bloody joking. What's she got them knitting needles for?"

Igor hadn't finished. He said the clothing was all in the mind, too. Just a question of status and conformity. He adjusted his silk pocket handkerchief.

"Remember we talked about Darren's clothes? And the people with dozens of suits and summer dresses? Over a ton of emissions a year from making and washing clothes?"

Joe nodded.

"Have you ever wondered why we buy all these clothes?"

"Every day. They cost me a bloody fortune, these kids always wanting new clothes. Me and Frank we made do with cast-offs and short trousers."

Kelly pouted. "But Dad, I can't go to school looking like someone out of Oliver Twist."

"Right, and I'm not going to play for the juniors in a pair of Billy's Boots, that's definite," said Darren.

"Billy's Boots?" asked Igor.

"From Roy of the Rovers. An old leather pair of football boots, that's what he means," explained Joe.

"So what if everyone else was wearing Oliver Twist kit? Or Billy's Boots?" asked the Professor.

"Well they wouldn't, would they?"

"But … imagine … what if they did?"

"Then it'd be all right, wouldn't it. That'd be cool."

"There you are, then," said the Professor. "The wardrobe is for the friends, not for Kelly. Boots are for cool. It's about conforming. If it wasn't for conforming, Granny Sugden's knitting would do."

"Well I wouldn't be seen dead in Granny Sugden's knitting," snapped Kelly.

"Come on, that's going a bit far," said Darren. "I mean…"

"But that's his point," began Doris. "If everyone wore their granny's knitting…"

"They'd look a right bunch of twats," said Joe.

"I can't believe it," said Doris. "I mean we spend hundreds of quid on clothes every year … and you're saying it's just conforming with what everyone else does?

"Isn't it the same with iPhones and playstations and Game Boys and big screen tellies?" she continued. "It's only because other people have them? We don't really need it, do we?"

"Flipping heck," said Joe. "That makes next Christmas easy."

Darren wasn't satisfied. "Come on Professor, we know all this. If anyone thinks about it for a minute, they know that all the stuff we buy isn't really necessary. It's just … well, we need the kicks, the novelty."

"I know, Darren. But remember the three Smarties."

"Yeah I know, but …"

The Professor rattled off a few more examples of things which were all in the mind: leaving lights on, not switching TVs off, filling the kettle full when you're just making a cup of instant, drinking endless hot drinks all day long.

Joe said he knew one. Those pillocks with leaf blowers.

"Ah yes, so-called gardeners shifting leaves around their gardens with the strategic intelligence of a small dog chasing its tail. An absurd technology and there are grave psychological problems there … but it's not significant for our analysis. Too small, I'm afraid."

Doris suggested the vacs for crumbs on the kitchen table and Joe recalled that their neighbour Gladys had bought an electronic pepper grinder the other day.

"All true, all true. But very small."

Igor stood up and stretched his legs. He located the second wine bottle, pulled out the cork and poured another glass. He savoured it.

"How much did that rob you, then, Prof?"

"Joe!" cried Doris.

"It was well worth the price. Not a nice thing to bet on a missed penalty at the Everton game, but still."

Then he moved on to transport claiming that many of our transport choices are just about habit and state of mind. He explained that we often think that it's quicker to go by car on short journeys. But by the time we've circled around looking for a parking space we could have walked there or taken public transport. We like the car because we're lazy – but what's laziness if it's not just a state of mind? Or sometimes we take the car just because we don't know that the number 25 bus will take us the same place. Not knowing that there's a number 25 bus – that's also just a state of mind.

He said when people compare how long a journey takes by car with how long it takes by train, they often simply forget that the time spent in the train is useful time while the time spent in the car is not. Forgetting, too, is all in the head.

Joe was half-impressed.

"My last and most important example is expense. People think expensive is something absolute – to do with mathematics and financial returns. Nothing of the sort. It's completely subjective. We can afford to spend £15,000 on a new kitchen, but not afford to spend £15,000 on a ground-source heat pump or top quality insulation.

"We can afford to spend ten pounds a week on cigarettes but not spend that much extra on fresh food. We can afford to spend £300 to fly to Prague for a stag weekend but not fit ultra low-energy light bulbs in every room for £50 all in. We can afford a Golf over a Fiat 500 – perhaps £6,000 more, but can't afford to sign up for green electricity which is £100 or so a year more expensive than normal electricity.

"Expensive is not what we can afford but what we want to afford. It's to do with our minds and not our pockets – psychology and not economics."

"So you're saying," asked Doris, "all the things we do in our daily lives, it's not because we really need to, but just because of what goes on in our heads?"

"Exactly, my dear!" said Igor. "We do these things just because of our psychology and not because we really have to. Not even because they make us any happier. All the reasons for causing emissions are rooted in our psychology. Interestingly I did not find any power stations among these reasons for emissions of greenhouse gases." He paused. "Now," he said, "Let's get off to the game."

7.   Routine

*A special case of behaviour which causes emissions is
the patternof routine and escapism.*

We snatched a one-nil win over Cerno More and got through to
the group stages with a draw in Varna. Back at home life was less
glamorous. We were trounced 5-1 by Chelsea and dumped out
of the Carling Cup by Middlesbrough. Then we scraped a draw
away to Liverpool. Joe swung from ecstasy to despondency and back
again. A look at the league table and he slumped into gloom.
Beneath us was only Hull and then humiliation and oblivion.

"You're up and down like a yoyo," remarked Igor. "How do can
you bear this? Does it go on for ever?"

Joe thought about it. "Well, that's the thing, it doesn't. Each season
you start again from scratch, don't you? The slate gets wiped clean,
you wipe away all the pain and disappointment. Three teams go
down and three come up from the lower league. But everyone else
just starts all over again with zero points."

"Excellent!" exclaimed the Professor. "It's the perfect solution to the
problem. In our lives we have our habits on the one hand and we
have novelty on the other. One is all about repetition and routine, the
other is all about change and escape. The more we have habits, the
more we need novelty. And both are traps for emissions – our bad
habits like eating meat and leaving the heating on, and our novelties
like buying new phones and clothes every season. Then even our
novelties become habits, so we have to have bigger and bigger things
to get excited about: we have to escalate our demands all the time.

"Now, I see with football you really are much cleverer. You
have your routine – the weekly matches – but each is 90 minutes
of novelty, because the matches are unpredictable and full of
excitement. And you avoid escalation – because you start again
from scratch each season. If only we could wipe the slate clean
with our habits and our novelties, our emissions would be a
fraction of what they are today."

8. Digging deeper

*Igor starts to delve even deeper into our minds to
understand the causes of emissions.*

Mid-September brought a cold and rainy spell – no sooner had
autumn come, winter was already waiting impatiently on the
touchline. On the night of the 17th we sat in the terraces in damp
claret scarves watching us host Beşiktaş in the first match of the group
stages of the Europa League. The floodlit pitch glowed emerald green
in the evening light. The stadium was packed and soggy.

We held out right up to the end of the first half, when Hološko
blasted Beşiktaş into the lead with a 32 yard screamer. Joe gritted
his teeth. He wouldn't even accept a half time beer. But our boys
must have had some roasting because the second half was fearsome.

We were watching the crusaders and the defenders of
Constantinople locked in a fight to the death. Crusaders in claret
and blue. Saracens in black and white stripes. St. George struck
in the 53rd minute – a free-kick whipped in from the left,
inswinging, missed by all, bounced in off the post. Then Beşiktaş
regained the lead with a low cross from the right bravely headed
in by a diving Nihat. St. George battled back, stubborn, inelegant
and gutsy – a corner in the eightieth minute was fumbled by
Rustu, and Paterson bundled it over the line. Still the men in
claret came, the Turkish goal was under constant siege, the grass
was red with the blood of heroes. With only moments remaining,
McDonald thundered a cannon ball from the edge of the box, it
cascaded off one man, clattered off another, and rolled past the
wrong-footed keeper into the Turkish goal.

We celebrated with two bottles of the best.

"Indeed, my friends, our heroes dug deep. It was a remarkable
display. It was … primaeval. Almost bestial."

After we'd said goodbye to Joe and Doris, the Professor and I
strolled through the streets of Brierfield.

"Unfortunately, my friend," he said to me. "I believe that we still haven't dug deep enough. Many emissions are the product of our psychology – our habits, our comforts, our tastes and preferences, our novelties and thrills – they're all things of the mind and things which can be changed.

"But there are deeper forces at work: the urge for fulfilment, the drive of ambition, the hunger for status, the compulsion to feel busy ... and then even more fundamental thoughts and attitudes which shape all our behaviour: the desire for physical things like nourishment, warmth, security, company, fun and rest; and the desire for psychological things like love, recognition, and fulfilment. And then," he added, "We want kicks. I need to speak to some more people."

This is what he discovered.

JOE'S MUM, MRS SUGDEN: Mrs Sugden had baked some scones and there was home-made jam. She poured out the tea in her best china. "It were all different in the old days ... none of this nonsense with families split up all the time. Mums and dads lived together. Family had tea together in the evening when dad got back from work. You see, dad went out to work and mum stayed at home and spent her time with children.

"She gave 'em plenty of attention and ... well, old-fashioned love. And it turned them out sorted and happy ... they didn't need to express themselves and discover themselves and all that fancy stuff like that ... and they didn't want a lot.

"Then something happened and it all started going wrong. It was hard to get a job or keep one down; lot of foreign stuff started coming in; values ... well, values went down the tube, didn't they? Load of immigrants, didn't know the ropes. A lot of rubbish on telly. Drugs and the like. And music. You know what I mean.

"Stan had trouble at work, and we didn't have a telly and he said look the Shaws are getting on, aren't they? Look at Mrs Shaw in her Triumph – the car, that is – he went on and on, so I started working ... and then children had less attention and started creating. I mean I was trying to do ten things at once, it were tough, and me and Stan started rowing ... You know, I got stressed up, and Stan got stressed and we started having rows, and then in the end we split up, and he went off with Mrs Shaw. Well, that didn't last long and she went back to Mr Shaw, she knew what side the bread was buttered on, and then Stan went off to Yorkshire. That was it. And we just prayed that the TV licensing van wouldn't come down our close. The kids were messed up and I couldn't spend the time with them... You can't imagine, Professor.

"Always creating they were, wanting attention and causing trouble at school ... their mum was out at work and Dad had buggered off to Beverley. They started nicking stuff – Nikes and what not to show off at school ... Well, they weren't going to make the same mistakes as their mum and dad and end up without enough money for a TV licence.

"Luckily they fell on their feet and I mean look at Frank... You haven't met him yet? He'll be back in the week ... But I mean Joe and Doris ... just when they realised that you have to work blooming hard to make ends meet, then credit cards came out and their kids could have everything they wanted.

"All they do is Boundary Mills and Accessorize and TK Maxx ... life's just spending on clothes and gadgets and new phones and iwhatsits ... all more clutter, all more energy, isn't it Professor? I don't think we've got any happier.

"Well, dear ... you can't go back, can you? I mean what would you do with all the people in advertising, and air-hostesses and car salesmen ... people selling you credit cards ... journalists and marketing this and that, and your banks and insurance ... your financial advisors and mortgage people ... your pilots and luggage carriers ... all your television personalities ... all them that make flat-screen tvs. I mean, what would happen to them all?"

ARTHUR HAKE, ZOOKEEPER: Mr Hake was partial to a pint, so they met in the Bridge. He wore a neat moustache but his tidy appearance couldn't disguise his faint whiff of animal. "I have been observing animals, including human beings, for over fifty years, and it's all very clear to me. All of us in God's kingdom

act in response to a number of visceral stimuli or urges: hunger, cold, fear, insecurity, and an overwhelming desire to mate. That's what nature has put in our genes. Just look at them come spring." He stroked his upper lip.

"Everything else we do is a function or derivative of this. Even hunger for political or social power is merely searching for an indirect route towards satisfying one or more of those primeval urges. Wealth – the accumulation of moneys – itself is just a way to get us closer to security and power and a trophy mate. When people are satisfied by regular eating and copulation their urge for power wanes. I have observed this myself in the gorilla's den. I mean strictly 'observed'.

"It's all down to psychology and physiology. These are the underpinnings of our behaviour and therefore of man-made emissions."

Mr Hake selected a crisp.

"Now, you can strip down our psychology into bits, just like you can strip down an engine. Some parts are more or less fixed – we physically need food and shelter – and there's nothing you can do about it. If I don't water my gibbons every day, I'll have all hell to pay. Yet there are other parts of our psychology which are more flexible, more susceptible to influence, more malleable. For example, the urge for power can be satisfied and made harmless through rigorous competition in sport. Material ambition can be sated through taking humble pleasure in bee-keeping. The collecting urge, through bird-watching or the taxonomy of postage stamps.

"A lot of the things we do which cause emissions of carbon dioxide – they're just something we've seized upon, something – often arbitrary – to be the vehicle for satisfying an underlying need. For example, we drive a large red car to project power and sexual prowess. We build a big house to satisfy a sense of security or to project power. We shop 'til we drop because our hunting instinct is otherwise unsatisfied.

"We like to flaunt our status and catch up with those who already have status. But we don't have coloured bottoms like mandarins, outrageous designs on our noses like toucans, elitist hairdos like lyre birds, or the even ability to conjour up bad smells like a skunk. Therefore to communicate status, excitement, achievement, or shock we have to purchase it from Top Shop, Dolce & Gabbana, Hermes or Johnny Choo.

"The things our minds hit upon to satisfy these urges are largely arbitrary. We follow the fashion set by others. Those fashions are the product of whims. Few characteristics of material goods represent fundamentals. Even size is not a fundamental since in some cases large objects project power – such as Hummers, while in other cases, small objects – such as mobile phones. As to the view of my wife, Matilda, you will have to ask her," he added pertly.

"This is at the heart of tackling climate change. Our ambition and our aspirations can all be influenced and channelled into far less harmful areas. Our choice of material comforts is arbitrary. Thus it's susceptible to challenge, refutation and change."

He knocked down the rest of his pint. Then, carefully removing a llama's hair from his upper lip, he stood up, thanked the Professor for the drink, and returned to the cages.

DR. ALBRECHT DOPPELGANGER-FISCH, AMATEUR NEURO-SURGEON: Doppelganger-Fisch was clearing up after a busy day's work, sharpening scalpels and oiling forceps as he talked to the Professor. "The economy is nothing but the organised discipline of dispelling boredom. Beyond the supply of food, shelter and bedding materials, its task is to generate flows of chemicals in our brains through creating aspirations and then providing, at a price, the means to achieve those aspirations.

"You see, boredom is tackled by the creation of flows of dopamine, adrenalin, serotonin, endorphins, and other chemicals in our brain and body, to give senses of achievement, comfort, pleasure, challenge, satisfaction, contentment, wellbeing and ecstasy. And then there's the small question of testosterone, of course.

"It doesn't matter whether boredom is dispelled by flying to casinos in Monaco or by a round of backgammon in the local pub. The same chemicals are at work.

"Consider that the evolution of our brain coincided with perfecting the art of the hunter-gatherer. To satisfy our basic needs we were forced to exercise all our faculties – alertness, speed, strength, calculation, and aggression – and we experienced in plenty the associated emotions of the excitement, fear, daring, hope, and achievement. We lived short and dangerous lives, but fulfilled ones. Once our needs were met we could be idle and drink the fermented juice of fruits or dance or play on drums.

"Today the only hunting is for a parking space in Tesco. Unless you're a bottle-blonde reversing an SUV while speaking on the mobile phone, that doesn't represent a challenge. The only foraging is for the discounted avocados in the vegetable shelves. Food comes to us as effortlessly as it does to our cats. Tesco is a vast, self-refilling tin of Whiskas.

"Can we generate chemical flows in our brains in a low carbon way? Could we escape boredom by channelling our enthusiasm and aspirations into non-polluting activities? I really don't know, Igor. I am only an amateur. But since our choice of activities is arbitrary or often only loosely based on the fundamental features of those activities, it should be possible."

HYRAM J.J. SILVESTER III, ADVERTISING EXECUTIVE: "My job is very simple: to get people to spend money. The sure-fire way of doing that is to make them want things. So I have to make them feel inadequate, missing something. If someone is happy they won't buy. So I have to destabilise them, upset them, undermine their state of wellbeing, to get them to spend.

"The great thing is this: people are stupid. They fall for it. They really believe it.

"That is what I do all day and everyday. It's what I love doing. It's what I am the best at. That's why we're sitting here in the Ritz over a bottle of Pétrus. Don't you forget that.

"Of course we have the right people on our side. Business, finance, and the law-makers. And the media. We've created the perfect storm: business needs us because without us they couldn't sell; media needs us because without us they wouldn't have advertising revenue; and politicians need us because we keep the economy spinning round.

"We make you want to sit on sunny beaches or slide down snowy mountains. I couldn't think of anything worse than standing hot and sweaty with the kids for hours in immigration or sitting among dogshit and cigarette butts and Germans on the beach – but that doesn't go into the dream. We make you want to drink fizzy drinks. Tastes shit and gives you diabetes, but we keep that bit out.

"The power, the feeling of power, it's incredible. We make people dream of granite tops in their kitchens, but they know they can't afford it. So we sell them glorious quantities of bank debt, so they can afford it. The ads don't tell them they'll have to work harder to meet the repayments and so'll have less time to spend with their children resulting in stress-related illness and family breakdown.

"But once we have them at the stage of family rows and screaming kids, we can sell them another cycle of escapism – comfort food and plastic and electronic tat for all the family. And of course the whole medical thing."

9.   Mirages

*The Professor argues that some of our fundamental motivations –
which lead to emissions – are delusions which could be corrected.*

When we all got together again, Igor started to explain the conclusions he'd drawn from his field work. "Are you telling us that it's all in the mind?" asked Joe. "That all the things we do are just … chasing after … those things in the desert?"

"Camels?" asked Doris. "Arabs?" said Darren. "Palm Trees?" said Kelly. But Darren said you can't chase a tree.

"Mirages. Chasing after mirages."

"That's deep, Joe," said Doris.

The Professor said there's nothing new in this. People have known for ages that money doesn't make us happy. "Look at Joe's friend Betty from Chorley who won the pools and then threw herself under a train. The cutting edge of sustainable transport."

There was uncomfortable quiet. The Professor hurried on. There was his mum's childhood in the Ukraine when they were happy but didn't have a penny. Then there are the Americans, the richest and biggest economy in the world, and they spend all their time talking to therapists. And how since the 1970s the UK has got richer and richer and sadder and sadder. For some reason we chase the delusion that wealth makes us happy, even though we know it doesn't work. We end up dissatisfied because the way to get people to want to be richer is to make them unhappy with what they have.

The Professor said that emissions of greenhouse gases are related to getting richer. The richer we are, the more we consume, and the more we consume, the higher our emissions are. And yet so many of the reasons we give for doing carbon intensive things can also be dismissed as delusions.

"Psychology is at the heart of many things which we do. Flawed psychology."

"He's right about psychology," said Joe. "For a start Alexander wouldn't have missed that penalty on Sunday if he had his psychology right. And Jensen wouldn't have fumbled the cross that led to their first goal. And we wouldn't have just given up, would we?" Joe was still pissed off at the game which followed our success in Europe. In the muddiness of the Premier League we lost two-nil at home to Nottingham Forest. A dismal result on a dismal, rainy day.

"Psychology can close the gap," Igor said. "Closing the gap could mean survival. Our everyday routine – it's driven by our minds and needn't be like that. The things we do to try to become happier – the things we buy, the things we covet, the things which we work late for, the things we drive hundreds of miles for, the things we borrow for."

"Our minds get cranky. They're twisted up by advertising, by what we see on television, by how we're brought up, by stresses of urban living, by cut-throat competition, by the high-street, by the pressure of our peer group … and they're starved of the healing stimuli of nature. It mixes us up."

"Most of our economy is dedicated to the unsuccessful pursuit of happiness. All it gives us is pollution and unhappiness. If only there were policies which would take advantage of this … if our politicians had the courage …" he tailed off. "But I suppose you still need to keep warm, and you need to eat. Other than that …"

"That doesn't sound much fun," said Doris. "All we'd do is eat and keep warm. What about all the nice things in life? You'd get bored if that's all you did."

"How about eating, keeping warm, and watching Burnley?"

asked the Professor.

"Could we have beer?" asked Joe.

"Of course. And wine."

"And would there be …"

"Absolutely. Plenty of it."

"Then perhaps it's not so bad after all," said Joe.

10. Why it's hard to cut emissions

Igor moves away from what causes emissions, to why it's hard to cut them. He looks at the variety of things which stop us changing our behaviour.

The following Saturday we hiked down to Craven Cottage in London and were rewarded with a win against Fulham. Eagles's free-kick swirled magically in the wind, took a deflection and bamboozled their keeper Mark Schwartzer.

At the Professor's suggestion we made a weekend of the visit to London.

As we surveyed London from the Eye, we discussed why it was so difficult for Joe and Doris to switch to a low-carbon lifestyle; what was preventing them from making the shift.

We started with their holiday. Joe and Doris like a week in Benidorm in the summer. What if they took the train instead of flying? It would take 24 hours just to get to Barcelona by train. "Barcelona? Is that near Benidorm?" exclaimed Joe. "We could just stop there, couldn't we? Spend a week at the Camp Nou. Better than the camp site." He roared with laughter.

Doris put her foot down: "I'm not taking the bloody train to Barcelona. It's hot and smelly and it'd undo any benefit of the holiday." The Professor suggested having a holiday in England.

But Doris said that Scarborough's not quite the same.

"Well, you can get the same food in Scarborough as you can in Benidorm. Same pub chains, too," said Darren.

"What about the sun?" she asked.

"Gives you bloody skin cancer anyway. Look at Joan at Number 13. She'll be dead in six months," said Joe.

"Water sports?"

"Dad knackered his knee on the jet-ski in Benidorm," said Darren.

But Doris had standards to be met. "Not go to Benidorm? What would Irene say? They were in Dominica last year. Going to Zanzibar next. And you tell me you want to spend a week in Scarborough? I could never live it down."

"But Chopper said Dominica was nine hours in a frigging plane, turbulence like the big dipper at Blackpool pleasure beach, a fat bastard with bad breath snoring next to him most of the time, puts his bloody arm over his shoulder. You don't need to go to bloody Africa for that, do you? There's that club in Blackburn."

The next destination was the shops in Oxford Street. Kelly needed a new iPod. Darren was looking for some new Nike trainers. Doris took a fancy to a blender in John Lewis. Then Kelly picked up some jeans in Next and Darren found a new hoody in Primark. Not to be outdone, Doris bought three t-shirts and a pair of joggers in a sale off Regent Street.

"They bloody think it grows on trees," groaned Joe. "We'd have been better off going to the zoo. They'll only shove it under the bed after six months and Doris'll give it to Cancer Care a year later."

It was scarcely possible to breathe on the tube from Oxford Circus up to Baker Street. Joe said this was why he takes the car. It goes when you want it to and it doesn't have other people in.

"But what if they had comfortable buses with lots of space so you don't get your nose stuck under other peoples' armpits?" I asked.

"Not much I can do about the bus service, is there?" said Joe.

"You could write to your MP, Dad."

"You could drive an electric car," I suggested.

"Yeah, actually I rang the Ford dealer, but they said they didn't have one expected in until 2015."

Doris brought up car-sharing. "Kevin goes to Barrowford most Thursdays and Fridays. And you could go with Gav on Mondays."

"What a good idea," enthused the Professor. "That would cut emissions by a good 30%. A fine start."

"What, I'm going to call Kevin and say, 'Hey, Kev can you give me a lift on Thursdays and Saturdays?' 'What, lost your license, mate?' says Kevin. 'No it's just … I'm er trying to reduce my emissions.' What's Kevin going to say? He'll think I'm a right fairy. And what do I do when the business moves to Keighley?" Igor confirmed that the forty minute car ride from Burnley to Keighley would take over an hour and a half on the bus.

"You could take that job in Reedley for a fitter. That's cycling distance," said Doris.

"Love, I'm not going to change me job just to be closer to home. Even if it's the same amount of money."

We arrived at Baker Street and went to see David Beckham and Lewis Hamilton at Madame Tussauds.

Kelly was admiring Lance Armstrong.

"You could be at work on a bike in no time," said Joe to Doris.

"Who's going to drop Kelly off at school?"

"Get one of them double bikes," said Joe, inspecting Indian cricketer, Sachin Tendulkar. "Tan… something."

"Tandoori?" asked Darren.

"Tandem."

"Get away," said Kelly. "I'd look a right twat turning up at school on a tandem."

"What's new," said Darren. "Anyway, it would do you good."

"It's not safe, is it," said Doris. "Biking through Burnley in rush-hour traffic?"

"And you'd be fairly sticky when you got to work. Do they have showers?" said Joe.

"You see," pronounced Igor. "There's a fascinating complexity at work. We're looking at one small question – how Joe and Doris can cut the emissions from getting to work. And even this small question is wrapped up in issues such as public services, the cost and availability of technology, design of roads, social conventions, corporate policies, design of offices, work-life balance, driving culture, perceptions of road safety and attitudes to convenience, comfort, exercising, and personal hygiene."

Then Igor insisted on visiting Kew Gardens. We ambled through the palm house.

"I could handle living here," said Doris. She folded her coat over her arm and took off her jumper.

When the girls had disappeared to the gift shop, Darren said to the Professor: "You know, mum was really pissed off the other day when you said she could only have the heating on 15°C. What did she have the boob job for, if she can't be in a t-shirt all the time?"

"Boob job! £5,000 which could have been spent on triple glazing the sitting room," blustered Igor. "All right, let us say that for 'medical' reasons, the minimum temperature which Joe and Doris can tolerate in their house is 17°C. A compromise between goosepimples and boobs. This is the work we need done." Igor showed us a page in his notebook. It was a shopping list of all the work to be done on Joe's house to cut energy demand to a few quid a month. Joe was quiet for a moment as he looked over the list. He shook his head.

"You can forget about this," he said. "Even if I had all the money in the world, I couldn't get this done. Here's that heat pump whatever it is. Do they do them in Comet?"

"You could try Mr Gambino of Gambino and Sons, Heating Fitters in Accrington," suggested Igor, dipping into his note book.

"You won't get Dad calling anyone in Accrington," protested Darren.

"Then if you call Mr Pepperoni of Pepperoni and sons, Heating and Plumbing Services, Chorley, Lancs."

"And how do I know if they're honest and reliable and charge a fair price? Then who's going to clear out the living room before they come? And how do you know when they're going to turn up? Who's going to sit at home while they're here to watch they don't nick owt, tidy up after them? It's a week off for me. And then there's the plumbers, the solar panel fitters, the glaziers, the electricians, the cavity wall people, and the roof insulators. It's months and months."

"Couldn't Uncle Frank help, dad? He's in the building trade. And he can negotiate the arm off a professional boxer. He'll see that there's no crap from the workmen," said Darren.

"Thank God for Uncle Frank," said the Professor. "Sounds like just the man we need." He paused. "If only everyone had an Uncle Frank."

"What else have you got on the list?" said Joe, taking the note book from the Professor. "Washing machine, fridge-freezer … How the hell would I know what the most energy efficient washing machine is? I haven't got time for all this.

"There's something else," he added. "Mum."

"Mum?"

"Doris. Think of the mess all these invasions will create. Dust and clutter. We'll have to move the furniture out and cover everything with sheets. All that hoovering and dusting afterwards. And think of the moaning. Women can predict all sorts of details which we don't think about. I can't face it."

11. More on why it's hard

Some more on why it's hard to change our behaviour.

The next day we went to the Science Museum. Darren felt unwell after the space-age ride and so we gave Glimpses of Medical History a miss and went down to the café.

Joe tested Igor on his knowledge of Burnley, its history, its players, its tactics and its opponents. Joe was amazed at how he could pack away the facts and figures. Doris laughed that he'd soon know more than Joe but Joe didn't laugh at that.

After finishing off the 1970s team lists and league finishing positions, the Professor turned to emissions. "Knowledge is certainly one of the reasons why it's hard to cut emissions – to live without emissions in your every day life you have to know bus and train timetables; where to get organic food from; who stocks LED light bulbs; how to play card games or knit or play a musical instrument; how to read an electricity bill; how to grow tomatoes; where there are clean beaches in England; how to cook interesting food from things without legs."

"Well we haven't time for all that, have we?" said Doris. "It's enough remembering when Darren's got training and Kelly's got dance classes. How can you learn all that stuff?"

"I quite agree," said the Professor. "It requires a level of dedication that only very few people have. For the rest of us, life is complicated enough as it is." With that he took a bite of toasted tea-cake and fell into a reverie.

It wasn't until the tour of the Tower of London that the Professor started up again. "Even if you did have that dedication," said the Professor, "there are constraints around you which you can't escape, however much you try. Just as there are games which Burnley can't win even with the greatest of concentration and strongest will-power."

"Are you sure about that, Prof?" Joe reminded him of Burnley's win over Manchester United in the previous season.

"Glorious events will occur from time to time. But to win week in, week out … that is something else. Things will happen which are outside their control. The strength of others – that we can't control. We have no influence over whether PSV hire Drogba or not."

"Many things are outside our control. And there are many reasons for this," continued the Professor. "Even if you want to cycle or take the train, you or Joe can't create safe cycle paths for yourselves. You can't individually improve the bus routes or add extra buses, or make the buses more suitable for carrying shopping in. You can't build new train lines. Doris might want to cycle to work but she can't install showers at work. If Mr Trew moves to Keighley, Joe has to follow.

"However much you spend on insulating the bungalow, there's only so far you can go. A bungalow is inherently inefficient.

Wouldn't it be better if the whole close were knocked down and replaced with condominiums?"

Kelly and Darren giggled. "What's a condominium?" asked Darren.

"A condominium is a – "

"Ignore them, Professor," said Doris.

"Wouldn't it be better if we all lived in tower blocks? That'd be dead efficient," said Darren.

"Oh yeah, try getting a hundred families to agree on installing a new heating system!" said Joe. "We can't even agree among ourselves on stuff."

"Or if we were renting," said Doris. "You can't just go ahead and put in insulation or what not if you're renting. You don't have the rights to."

I didn't want to ask what they'd put in Doris' coffee: she was on fire.

"And we can't build our own personal nuclear power plant or wind farm. The technology doesn't work at that scale, does it, Professor?" added Darren.

"Indeed. And you can't just switch off your freezer, because you haven't time to buy fresh food every day or two. You can't cut the wasteful emissions from drinks chillers by not buying chilled water. You can't have wind farms put up if everyone around you opposes them."

"Might as well not bother, then, Professor, if there's nowt we can do about all this," said Joe glumly.

"It sometimes feels that way, my friend. The individual can only do so much. After that, irrespective of his wealth or enthusiasm, he bumps into brick walls.

"So … you're saying that however hard Burnley try … we just won't be able to win the league," asked Joe.

"Our parallel doesn't work in all cases, my friend. I believe that if Burnley show exceptional character and drive, then they can … er … at least avoid relegation. But today amateur excellence rarely reaps success, I'm sorry to say. We aren't billionaires, we don't have 'global appeal'. There are some things which we just can't do by ourselves. There are players whom we can't afford. In some matters we need help from above."

"What," asked Joe. "You mean from … God?"

"Him too. But I meant from those who direct this game … who influence the economy and society."

12. PSV Eindhoven

A rest from emissions.

"Just breathe it in," said Joe. He was explaining to the Professor the unexplainable. The thrill of a tie in Europe, the bright green pitch, the floodlights, the chill in the air. "Don't you feel it?" We were sat in the Philips Stadium soaking up the atmosphere.

Drogba and Dzsudzsák weren't on good form. Wade Elliott and Steven Fletcher fizzed at the front, and Edgar and Cort formed a stone wall which the Dutch couldn't penetrate. Then just before half time, PSV made a tactical switch. In the moments before Burnley could react, Affelay had made a one-two with Dzsudszák, split the Burnley defence, laid on Pieters, who lifted the ball over Brian Jensen.

At half time Joe joined the long queue for beer.

No sooner than the players came out, Mears punted the ball upfield. Elliott controlled it, nutmegged Toivonen, slipped past Maza's challenge, rounded the goalkeeper, and tucked in the equaliser.

The Burnley sector exploded with joy. We had the precious away goal.

A few minutes later Joe came back, laden with beer. He'd missed the goal and he'd got tomato ketchup from the hotdogs all over his Burnley away shirt. And one of the plastic cups had split and was leaking beer. But who cared? We roared for the Clarets, we roared for our country. Even the Professor had a confused grin on his face.

13. Uncle Frank and democracy

*Igor and Uncle Frank have a chat about democracy and
its role in emissions.*

"Our success story," said Doris.

Uncle Frank arrived in a BMW 335i Cabriolet. He wore gold
and advertised his undying love of Cilla, Tina and Burnley FC
on his upper body.

"Don't listen to him, Joe," said Frank. "You've worked hard for
that Golf. You deserve it. So use it. Africans and polar bears and
bloody bits of coral aren't your problem. Plus nothing you do will
make any difference. Look at the bloody Chinese."

Frank and the Professor were going to get on famously. I'd
warned Igor about Frank's small talk.

The Professor asked Frank if he voted.

"Me? Vote? Never. They're a bunch of tossers them MPs. My
vote won't make a big difference, will it?"

"Any children, Frank?" asked the Professor.

"Children? Not me. Noisy bastards. Me and Cilla decided we
weren't going to have children and we're not changing that."

"Any animals?"

"We had a pit-bull. Had to be put down. Too violent. But I'm
getting a new one."

"Do you have any hobbies?"

"Hobbies? What do you think I am, a bleeding boy scout? Me,
I work and I watch Burnley. That's your Uncle Frank."

"You've got Cilla, love," said Doris.

"Right, and Cilla. I forgot about her."

Frank looked down at his right forearm and gave the tattoo a
superstitious kiss.

"So who was Tina?" asked the Professor looking at the other
forearm. "Was that the pit bull?"

"Here, watch it!" spluttered Frank reddening. "She was my wife
before Cilla. The other two … well you can't see them from

where you are."

"Four wives and counting, but I think your real love is Burnley," said the Professor.

"I've had a season ticket for twenty three years. Hardly miss a game."

"Good heavens. But why do you bother going to the games? Isn't it sufficient to read the result in the newspaper the next day?" asked the Professor.

"Miss a game?" replied Frank astonished. "You can't miss a game if you love your club. You have to be there and support them."

"True, true. Not that it makes a difference really."

"What do you mean?" asked Frank.

"What's the average crowd at one of your games?"

"You'll get twenty-two thousand at a home game."

"Exactly, so it doesn't make a difference at all whether you go or not. Your voice would be indistinguishable. I'm sure the Clarets would do just as well if you didn't go."

"That's where you're wrong, Igor," said Frank. "Don't you know that one and one is two? Two and two is four. Four and four is eight. If everyone had your attitude, then there wouldn't be anyone at the match. Of course I make a difference – if I decide to go, and everyone else decides to go ... that's how you get a crowd. A crowd is just a large number of individual people."

"True, true. So the way to get a good crowd is if everyone individually agrees to go to the game," concluded the Professor. "But what would you do, Frank, if everyone said that they don't make a difference ... and in the end no-one came."

"I'd make it compulsory. We've had bad attendances in the past. You can't just hope for the best that everyone just turns up. If you want to guarantee a good crowd, you have to force them. If someone says they're a Burnley fan, then they should have to go to the matches."

"That's not very democratic," said the Professor.

Doris rolled her eyes. "Don't get him on to democracy, Professor."

It was too late. "Democracy? Biggest load of shite ever invented. Democracy is why this country's in the mess it's in. There's far too many layabouts in this country, ranting on about rights and democracy and lesbies and immigrants and all that shite. What this country needs is National Service and a good dose of common sense.

"You don't win wars with democracy, Igor. Nature's not a democracy.

"Look at your climate change stuff. You want our Joe to change his heating system and you want Doris to bike to the shops, and you want them to fit better insulation and you want them to buy less crap. Well you'll never get them to do that in a thousand years if you tell them it's all going to be easy and they can carry on like they want.

"If you want folk to take you seriously you have to tell them plain and simple it's bad news and everyone has to muck in. Folk don't listen to nicely-nicely. We didn't win the war by telling everyone that Hitler's army were a bunch of fairies. If you want folk to fight, you have to give them something to fight for. If you tell them it will be carry-on-Charlie they won't believe you and they won't put their shoulder in."

"We're beginning to see more and more eye to eye, Frank. But you still don't think that it's really your problem, do you?"

Frank shrugged his shoulders. All eyes were on him.

"Well it's not, is it? It's business and politicians what do all this, not me and Joe. We're just simple people."

The Professor tried a new tack. "Frank, you're a builder, with your own firm?"

"And proud of it," said Frank.

"And you build houses, yes? Why houses?"

"Well, I can sell them, can't I? People want houses. They need me to build them."

"He's very good, Professor. He built this one," beamed Doris. "Cup of tea anyone?" We were all gasping and she went to put the kettle on.

"Do you build passive houses, the ones which use scarcely any energy?" asked Igor.

"Course not. No-one would pay for that, would they? It's not my job to tell people what to order, is it?"

"Absolutely not. You have to do what the market says. That's a rule of business, no?" said the Professor.

"Right, Igor."

"So, Frank. What a business does is dictated by the customers. By me and you and Joe and whoever's the customer."

"Right," agreed Frank.

"So if we want to get business to cut their emissions, the customer has to be ready to pay for it and has to ask for it."

Frank mumbled something.

"So it's not the problem of business," reasoned the Professor. "It's the problem of the customer, after all."

"Well … it might be. But, as I say, the politicians should sort it out."

"They need your vote, Frank. And by the sound of it, they'll never get it."

"Well, if you believe in democracy, Igor. But I don't think you'll get folk following your ways in a democracy. It stands to reason."

Doris returned with a pot of tea and a plate of Hobnobs. We all felt it was time for a break.

Joe sipped his tea. "Well, it's the big one tomorrow. They'll be wanting their revenge."

Tomorrow was the home game against Manchester United. League champions four years on the trot. And we beat them in our first home game in the Premier League last season!

"Tell Sir Alex about democracy. I'll bet there's not a lot of it in the Old Trafford dressing room. They might earn a hundred grand a week, but they have to do exactly what he tells them. That's how to get stuff done." Frank had the final word.

14. The distribution of power

*It's also hard to cut emissions because of the distribution
of political power in society.*

By half-time it looked as though we might just do a repeat. We'd
nicked an early goal before United had got into their stride. But
they must have found their stride in the dressing room at half
time. After that we had to keep eleven men behind the ball. They
got their equaliser – a free-kick from Rooney – but no more.

We celebrated the point over at Joe's.

"What if Frank's right?" asked Doris. "What if it's not our
responsibility? What if it's the job for government to sort all this
out? I've enough bother with work and the kids and Joe and
keeping the house going. You can't expect me to be saving the
rainforests as well."

"Well you'd be no good in rainforests anyway, would you?"
said Joe.

"You mean because of the spiders?"

"No. You don't speak Spanish, do you?" explained Joe.

"Dad, they don't speak Spanish in Brazil," said Kelly. "They
speak Portuguese. Like Ronaldo."

"Ronaldo? Perhaps I will go after all," said Doris.

"He's half your age, love. Forget it."

"Plus he's scared of spiders," added Kelly.

"How do you know that?"

"She's just making it up, you daft twat," said Uncle Frank.

"He's not my type anyway, love," continued Doris. "Anyway, I
was trying to ask the Professor something. Now where was I? Ah
yes. What if Frank's right and it's the job of government to sort
this out?"

"He might be right," admitted the Professor.

"If he is right," said Joe, "how come they haven't sorted it
out already?"

"That, my friend, is a very good question. It's a question

about power in our society, who has it and how it's used," said the Professor.

"Here he goes," said Joe. "Where's that beer, love?"

The Professor rubbed his chin. "In our society a small number of people at the top wield exceptional power and influence. Politicians, journalists, businessmen and celebrities. Because we live in big cities and everyone has got TV, they can amplify their influence in ways unimaginable in earlier civilisations. Look at shopping."

The Professor paced the room. He explained that in the olden days we'd purchase our goods from many small shops and market stalls run by entrepreneurs. Those entrepreneurs operated on a small scale which limited their influence and profitability. Now we buy from Tesco and Asda. Most people employed there earn no more than the man at the cheese market and his apprentice. But through the benefits of scale a small few – the owners and top management – enjoy stupendous wealth. He said it was the same with the media – a handful of newspaper and television proprietors control the flow of information to tens of millions of people.

And then power companies, oil companies, cement companies and so forth. Economies of scale have put our society in the hands of a few, highly influential people. It's simply the natural result of economic competition … and a little help from the people at the top.

"You mean everything is run by just a few people?" asked Joe.

"But that's awful," began Doris. "It sounds like something out of the Godfather."

"We should do something about it. I mean, the people should be running the place, not the toffs," said Joe.

Frank laughed.

"Would they do any better?" said Igor. "It's not in our interests to fight the status quo. We could lose our jobs, our livelihood, our liberty, or even our lives. And we're too busy just making ends meet, to start rocking the boat."

"If these few people are so powerful," said Joe, "why don't they force through the changes which we want to happen? Why don't they just make all those things happen?"

The Professor shrugged. "Yes, you'd think that the richer someone is, the more able they are to make a difference. If you're rich, you can choose the job you do. You're clever enough not to be taken in by the television and advertising. You don't need to seek status. You can already buy anything you wish. You can pull out any time. You can walk away. You can afford it."

"This is what I've been saying," said Frank. "That's why it's their responsibility. There's not much we can do here in Burnley about bloody cycle lanes and aeroplanes."

Joe wanted to know why they aren't changing things if they can.

"Remember what our friend Doppelgänger-Fisch said. As humans we seek to stimulate the flow of chemicals in our brain. But as the brain becomes used to certain stimuli, the effect of those stimuli weaken. Therefore you seek even greater stimuli. So a rich person constantly feels the need to become even richer and even more successful and famous. He's on the same treadmill as we are, just a bit further up, but the road ahead is infinite. First it was a souped-up Golf, then a Ferrari. Then it was a yacht. Then a private jet. Then space flight. There's always someone to beat.

"The second thing is that just because you're wealthy, it doesn't mean that you're any more altruistic than anyone else."

"Altruistic?"

"Rich people are selfish buggers like the rest of us. That's what he means," said Frank. The Professor nodded. He said that the characteristics which make men and women successful in business and politics − drive, individualism, aggression, competitiveness, ruthlessness − don't go together with altruism. A businessman without these qualities must be either a genius or a saint − or exceptionally lucky. "Altruistic and successful people are rare."

"'Me first', that's my motto," added Frank. "Now, where's that booze you brought?" he asked me.

"Good idea, Frank," I said. "This is heavy going." I went over to the tea trolley and opened the bottle of Chilean wine I'd brought round. The Professor sniffed something about the carbon footprint of Chilean wine.

"Now where were we?" he asked as he leaned back in his armchair clearly enjoying the wine. "Mmm … but next time let me recommend you a Gigondas. It'll cost a little more, but will keep you within your emissions budget." Joe yawned.

"Now, reasons why the rich and influential aren't leaders of change," continued Igor. "It's strange really. Just the people who could make change are the ones who don't want to. If you think about it, it's incomprehensible. It's almost a kind of paradox, isn't it?" I said.

"Yes, you could call it a paradox. But to elevate such a commonplace observation to the status of a paradox…"

"You could call it the Rowbotham paradox."

"Don't try and flatter me. I'd never dream of claiming origination of something so self-evident which hundreds have observed before. We aren't here for academic glory, but to help our friends the penguins. And don't forget Stigler's law. But let me continue."

The Professor spoke about competition. "Imagine Tesco says they will stop selling meat and actively promote vegetarianism. What would happen?"

"Well, Sainsbury's would just carry on selling meat and mop up," said Joe.

"Competition – another reason, then, why the rich and powerful can't make change. And then, however rich and influential the managers of Tesco are, in the end they have to do what the stock market wants them to. Their real bosses are the traders and analysts. Everything depends on their share prices.

"If Mr Tesco comes out with a statement that he wants us to eat less meat – and he has a commercial vision to back it, you can bet that the financial community won't understand it. Remember the character of traders – they behave on instinct. Their emotional intelligence is stunted. They're aggressive and beast-like, acting on their expectation of how others like them will act. If they don't understand Mr Tesco's argument, they will sell his stock."

"So even though we think that Tesco rules the world," said

Doris, "you're saying that the financial people rule Tesco."

But Joe asked why someone like Mr Tesco would let himself get tangled up with traders and analysts.

Frank put his arm on Joe's shoulder. "Because he wants to grow the business faster. So he needs capital."

"But why did he want to grow the business so fast?" asked Joe.

The Professor smiled. "You see, we've come full circle. We're back to the illusions. Why become even richer than rich? So that we're remembered for a hundred years not just ten years? Or for a thousand years? It may make them happier for fleeting moments. They always had a choice of running a small show in Lancashire and being able to spend plenty of time with the children. But it wasn't enough."

"Much wants more," said Doris. "Frank said it. You can never be rich enough."

"Once you do your deal with the financial community, you can't walk away. You've lost your independence, however rich you are."

"Doesn't sound much fun to me," said Doris. "I'm glad you're like you are, love," she said turning to Joe. "I'm glad Mr Trew's business isn't on the stock exchange."

"You say that now," said Frank. "But you wait till I sell out." He sipped his Malbec with a satisfied smile. "And remember this," he added turning to the Professor. "If you want people to come up with new technologies, stuff to cut your emissions – if you want people to invest in that, and build businesses to supply the stuff – you'll need rich people. And you need to give people the chance to become richer. So don't think you can go round playing the bloody socialist revolutionary."

"Of course we need them," agreed the Professor. "We need them more than ever – my question is what does it take to get them to play ball, to use their skills and influence for the good and not just for themselves?" But before we could answer, he moved on to the press. "Remember lots of newspapers are owned by rich, influential, and bullying people. Perhaps they were taunted or bullied as small children because they were weak or socially awkward. They're conditioned to be winners at all costs, and to have the loudest voices. To receive unlimited, unconditional recognition."

Doris said she couldn't see what this had to do with greenhouse gases.

"Because the press's job is to keep the myth of consumerism and economic growth fresh … and that's what causes growth in greenhouse gas emissions."

Igor's point was something like this. The media industry requires that lots of people read their newspapers and see their television programmes. The politicians rely on the newspapers and television for getting their message to the people. Therefore politicians are scared to upset the press. They feed the press with sound-bites. Anything complicated or difficult is overlooked.

But Joe said he'd thought the press was just what the Professor needed. "Imagine the press said the right things, though. Imagine they got people on your side. Just think how powerful they could be."

"Of course they could," said the Professor, "but they know that no-one would read any of that. The boss doesn't want a better world. He doesn't want facts and subtlety and thoughtfulness. He wants sales.

"My friend, people with power and wealth are beholden to the press, and the press is beholden to the people in power. It's the bond which keeps the status quo."

"So what can we do?" asked Joe.

"Patience, my friend. Patience." He smiled and emptied his glass. We turned to higher topics of set-pieces and the offside trap.

15. Instincts

Can we rely on people's instinct for self-preservation?

"It's hopeless, isn't it?" said Joe to me looking up from his Sun. We'd just got back from Blackburn where they'd beaten us two goals to nothing. Joe always buried himself in the Sun when there was no hope.

"You mean the Blackburn game?" piped up Igor.

Joe said nothing. The Blackburn game was something special. Losing it hurt double. He slumped back behind his newspaper.

Doris explained to Igor that Joe had it bad. "Look at him behind that paper, like an ostrich."

Igor sipped his tea. "Yes, our instincts are funny things. Take climate change, for example. If it's really such a big problem, why don't people do something about it by themselves, without needing the government to bother us with more laws and red-tape? Why don't we do something about it because it's in our self-interest?"

"If it gets bad enough, they will do, won't they? In the end, I mean," said Doris.

The Professor shook his head. "In the end they might, but the end is exactly what we want to avoid. Our nature is not equipped to respond to threats like this."

"Actually," began Joe, raising his voice. "I did mean the Blackburn game, but…" He sensed it was too late.

"Man is not naturally thoughtful," continued the Professor. "Our nature is to be aggressive and quarrelsome. We're ruthless and competitive and nasty and cruel and impatient. Sometimes we might be kind, but like a wildflower growing through the tarmac in an Asda carpark, it's an exception.

"Before we had technology, when our population was small, these tough qualities were necessary for survival. They became hard-wired in our genes by way of evolution. Aggressive men managed to win the best – the best birds, as you might say – and have children. Softies were pushed by the wayside. And these traits remain the principal defining characteristics of man."

"Like Frank," agreed Joe. He let his paper slip down onto the floor and drank up his tea.

The Professor chuckled. He explained that the instincts which we evolved were good at guiding our behaviour in small packs as hunter-gatherers in thinly populated territories: aggression, competitiveness, selfishness, perseverance, and so forth. The trouble was that they didn't work for us any more.

Doris wanted to know why.

"Unfortunately we haven't evolved a biological ability to sense environmental damage. Our noses don't sense levels of carbon dioxide in the air. We can't taste pollutants in water downstream from us. We don't feel pain when a tree is cut down, even if the removal of the forest is going to cause floods which will wash away our village. You see, nature's cry doesn't reach us. We're deaf to it. Apart from the smell of excrement on one's hand, even for the most simple forms of pollution there's no natural warning to us that we should stop it. Even with short-term environmental problems."

"So our instincts aren't any good for sorting out this problem," said Doris.

"Exactly, our evolution made us good at some things. But completely neglected other things. Our instincts serve us well in times of shortage but poorly in times of plenty. And some of those instincts serve us terribly when we want to cut emissions."

"Like?" asked Joe.

"Think about eating meat. It tastes good and comforting. In times of shortage to get a chunk of fatty meat down you is important, so we evolved a great sense of satisfaction from eating meat. With central heating and indoor living you don't need all that fat. But our instinct still makes us want to eat meat, and it happens to be the most emissions-intensive form of food."

Joe said Igor should give meat a rest.

"Then look at laziness, that feeling that you want to avoid effort. It's useful if you need to conserve energy. But it's disastrous if you have no shortage of food, because you end up not getting enough exercise. And it means we use mechanised energy to do work for us."

"Flipping heck," said Joe, suddenly bright-eyed. "Could that be why they invented football? I mean, so you can get exercise even when you don't need to go out hunting anymore?"

The Professor thought for a moment, then burst out with delight. "Of course! Football replaces hunting in times of plenty. And it replaces war in times of peace. Joe! Fine work!" Joe tried to give the Prof a high five, but Igor wasn't used to that kind of thing and made a mess of it. Doris said proudly: "He's not that daft, is he?"

"Not at all. We used to spend time fighting each other for resources or to get the best mate. Fighting speeded up the survival of the fittest because it was the strongest who got the best mate. But nowadays we don't need to fight, and people feel restless and itchy because they can't satisfy that instinct. This explains football. But unfortunately not everyone turns to football. Some devote themselves to self-fulfilment, ambition, chasing illusions, and so forth. And that creates our high-energy society."

"So nature isn't that clever is it? It only half-finished the job with us," remarked Doris.

"And ended up creating its own worst enemy," added Joe.

"But there must be some good instincts we have," said Doris. "What about mums looking after children?"

"Of course," agreed Igor. "We have many good instincts, too. But for our long-term interests, we have no instincts whatsoever."

"That's what Frank says," said Doris. "There's no long term."

"He's right. There's no long term. There's just the next game," said Joe. He picked up the remote and flicked on the TV.

16. Culture

If not instincts, can culture help us cut emissions?

A bit later Joe said: "So instincts won't help us. Funny that, really. It's the same with the team. If instincts were enough we wouldn't need to train, would we? And we wouldn't need tactics or a manager. They'd just play how they liked. That's when a team falls apart. You can have a team of instinctive superstars, but they can get beaten by a team of worse players with better tactics. Look at Germany game in the World Cup. Bunch of kids beating our lot on 100 grand a week."

Igor continued. "Thousands of years ago, hunters became so good at chasing after animals and killing them, that our populations grew, and there was less food to go round. We developed new ways of living and started farming and built villages and towns. In these bigger groups it turned out to be better to cooperate and discuss things and plan ahead and not beat each other up all the time. People developed new rules for survival in bigger groups. Sometimes those rules were brought from the sky in the form of religion. Sometimes they were unspoken codes and conventions, other times the rules were explicit codes like laws. All these rules kept us on the straight and narrow, and protected us from our own worst, selfish, self-destructive nature.

"The rules of our culture sat on top of our instincts and we worked out how to behave in a way which balanced the interests of the individual with the interests of the group. Joe?"

Joe had started flicking through the channels again. He stopped guiltily on National Geographic.

"I'm sorry, Professor, but it's a bit boring."

"But we're getting to the most important part!"

"Well … I'll try. But what've cavemen to do with it?"

"It's about how we avoid being cavemen. It's about how we move on from playing hopeful long balls into the box."

Joe looked up. "Hey – once I read that a hundred years ago when they started football, it was like everyone just dribbled as far as they could until they got into trouble. Then someone invented passing. It completely changed the game. That's culture, isn't it?"

"Exactly! This is about how we develop the thinking game, the passing game, the one which helps us survive in the Premiership. We are still like cavemen in most respects. We have a messy core of hard-wired instincts – for surviving in the rugged conditions of the Championship. Culture wraps around that and makes us into Premiership players."

"Just don't mention culture when Frank's here," warned Joe. "He says it's posh farts in tights."

"By culture I mean the rules about how men should live and work together. Or how they play together. The trouble is these rules also evolved before we knew how man could damage nature."

The Professor explained that our religions couldn't have imagined that we would develop technologies which can overpower nature. If they had, the Ten Commandments would start with 'Thou shalt not use a chain saw.' They wouldn't have placed man at the centre of nature, but on the edge. Near the rats and the foxes.

"There are some exceptions. From earliest times the wiser thinkers figured out our effect on the environment and exhorted us to protect trees and animals. And the wisest societies on the planet today – the forest dwellers of the Amazon – practice deliberate restraint because they know the limits of the nature around them. But in general, our rules are about how man should relate to man, not how man should relate to other animals or nature.

"This is the gap in our armour for survival. Where nature responds immediately – bad smells – our instincts help us. Where nature gives a response within a few months or years or so – in things like raising children, education, agriculture – our culture helps us get things right. But in things where nature doesn't give a response for decades or centuries neither instinct nor culture are there to help us."

"Like at Monday morning training you don't run like the clappers to win the league next May. You run like the clappers 'cos the gaffer's kicking your arse," said Joe.

We tried to understand Joe's explanation.

"Look," Joe tried again. "If I dropped a stone down a very deep well … if the stone only lands after I was dead, who'd give a monkey's about the frog at the bottom?"

"Very apt, my friend. Trying to cut greenhouse gases goes against our instincts – we've not evolved to handle long-term threats. So the politics which say people should do as they feel are unlikely to help us. Intervention is sadly necessary."

"Right. Like Darren needs a clip round the ear because he'll never learn to tidy his room by himself," said Doris.

"And remember we've very little time to figure things out. If you're bottom of the league at Christmas, you can't rely on eleven players working out new tactics by themselves by trial and error. The manager needs to intervene."

"Logical," agreed Joe.

Then Igor frowned. "Unfortunately, the world is going in quite the wrong direction. Instead of strengthening our culture with rules to protect nature, we are weakening it. The most powerful societies aren't those with the strongest culture and the most educated people. They are those who exploit technology the most ruthlessly. The characteristics you need to exploit technology – creativity, competitiveness, greed – are the product of our individual instincts not our culture. So to be even more effective at exploiting technology, we actually have to water down the cultural rules which we once created to help us."

"But don't we need to be good at technology?" interrupted Joe. "I mean what about medicine? And what about all the technology you need to cut carbon emissions? How would we be where we are without being good at exploiting technology? I mean, if we were all like little mice, then we wouldn't invent the mousetrap, would we?"

We took some time to work through Joe's example.

"Of course we need technology. I wouldn't live to 80 without technology," smiled the Professor. "I'm not complaining about technology. I'm complaining about unwinding our culture. The more freedom we have, the more we unravel our culture and its rules, and the more we revert to instinct."

"You mean we're like turning football into British Bulldog. Fuck the ball, fuck the goal, fuck the rules, fuck the skill, just get the fuck over to the other end of the pitch. Ouch! That hurt!"

Doris was livid. "Joe Sugden. If I've told you once to watch your tongue I've told you a hundred times. Now say that again nicely."

Joe hesitated.

"Or there'll be no beer tonight," warned Doris.

"Perhaps I could oblige," said Igor. "The cause of freedom and self-fulfilment has undermined our sense of society; our emphasis on short-term self-interest has made it even harder to tackle our long-term problems."

Joe was about to say something more about British Bulldog, but Igor wasn't done. He was worked into quite a lather. "We have a desperate urge for self-fulfilment but everything we do to satisfy that urge means using more and more energy. Just when we need to go in the other direction. We need self-restraint, modesty, quietness, slowness and a strong sense of community." Then he finished and drained his glass, wiping his brow.

"If I follow what you're saying, Professor, it's like you've got the FA Cup final coming up and the Gaffer cancels training and takes them all out for a piss-up."

"Simply said, yes."

"Then it won't work, will it?"

"Mind," laughed Doris. "If you could put that kind of passion into supporting Burnley, Professor … we might even have a chance against Lazio on Thursday."

17. Values

How the values which hold together society stop us from cutting emissions.

Igor still didn't get it. He understood football intellectually now. He could name formations and even recognise them being played out on the pitch. But he couldn't see why. He'd made an effort to learn some chants for the Lazio game but he sounded as passionate as if he were reading out the Pendle area bus timetable.

"No goals, and still terribly exciting?" asked the Professor. This was beyond him.

"Of course it's more fun if there are goals, but holding out for a nil-nil draw against Lazio! You've got Rocchi, Zarate, Ledesma and Pandeva blasting away at your goal, and our lads soaking it all up."

The next Sunday we got soundly beaten by Hull at the KC Stadium. Igor had been at the game but he was perplexed that his predictive model had been confounded. "I can only blame it on myself. I completely overlooked the scent of fish in the air. The high potassium levels in the smell of fish blocked receptors in our goalkeeper's brain, leading to a drop in confidence." Joe said that they should ban playing in towns with fishing traditions like they banned playing in cities over 3,000 metres.

We popped in to see Joe's dad in Beverley on the way back. We had a dreary cup of tea in the sheltered housing and listened to him griping about the rain and life at the bottom of the league and the tragedy of a Lancastrian ending up in East Yorkshire.

Perhaps it was something inherited – Joe continued moaning all the way home, through the front door and into the living room that evening. "It's always the same. We punch above our weight in Europe, but when we get back home something goes wrong. I don't know why we don't just buy a decent team. Why can't we just find a rich owner, like some Arab or a Russian, even a Chinaman, and he can buy some decent players for a change?" Joe grumbled.

"Come on, my friend. You have to lose sometimes. The probability of winning every game in the season is approximately 4 in a trillion assuming you start each game with equal odds. Even if you purchased the entire Barcelona team and went into each game with a 75% probability of winning, still the chance of winning every game is 50,000 to 1."

Joe tried to explain the agony of losing. "It's a kind of grief. It's a pain in your legs. It crushes your chest. It's – like you just know it's going to happen before it does."

"A sense of inevitability, irrevocability, foreboding. A dark, daunting sense of powerlessness against imminent irreversibility?" suggested the Professor.

"Got it," said Joe. "It's that shite feeling you can't unwind the past. Hull's goals – they're facts. Facts. You can't change it. It's happened. Nowt you can do. Tough luck. If you don't love Burnley, you'll never feel it. You have to love them. Watching them lose, it's horrible. There's nothing you can do except scream your head off."

"I do know what you mean by that the sense of powerlessness, Joe," agreed Igor. "I also sometimes ask 'Why don't we just ...?' 'Why don't we just ban big cars or walk-in freezers?' or 'Why don't we just tell people they have to install proper insulation?' or 'Why don't we just impose sanctions on such and such a country if it won't agree to cut emissions?' We sometimes wish that our political leaders were dictators, but good ones, on our side.

"We can't 'just' do things because the power of politicians is constrained by the constitution or by common agreement based on centuries of learning and experience. Deeper down, these arrangements reflect values which hold together our civilisation: liberty, rights, justice, peace, equality, dignity, compassion – and television.

"Often these values stand in direct conflict with the need to act firmly on cutting greenhouse gas emissions," continued the Professor.

"We don't force people to stop doing things because it conflicts with their liberty, their freedom to do what they want. Travel is a right. Eating meat is a right. Overheating your house is a right. Having single glazed windows in a town house is a right. Ambition is a right. Everything is a God-given right, and we can't 'just' take it away, whatever the cost to the planet.

"We don't threaten force on directors of energy companies. We don't jail advertisers for promoting energy intensive goods. It would conflict with our sense of justice.

"We don't invade countries because we disagree with their energy policy, since we would be breaking the peace between nations."

"Even if we're sure we'd win," said Darren.

"We're afraid of taxing electricity because we don't wish to harm the poor.

"We take our compassion to such an extreme that it's more a cruelty than a kindness. We can no longer distinguish between compassion and indulgence." Sometimes Igor got into these poetic moods. I kicked his foot.

"We're so scared to infringe on the liberty of the free market, that we readily excuse the suffering of its victims." Igor finished off.

"All he's saying, Dad, is we're so worried about being fair to everyone that we've lost sight of what's important," said Darren.

"But you can't just go round willy-nilly doing what you want," said Doris. "They're there for a reason them values. It's cos of mistakes we've made in the past. You can't just chuck them all in the bin."

"Well, it's bloody daft being nice to everyone if they're screwing up Darren and Kelly's future," said Joe. "Couldn't you like just ignore them for a bit?"

"Sometimes politicians do override these values in the common interest. But it's hard to do this, and the press and opposition politicians will attack them ruthlessly, unless it plays to the mood of the people," said Igor.

"What do you mean by the mood of the people, Professor?" asked Doris.

"When the people get energised about something, when they feel something's very important, then they're prepared to make sacrifices and allow the politicians to override the common values. Tough security laws in the US were made possible by the mood of anger and fear which followed the 9/11 attacks in New York. Governments get carte blanche to slaughter millions of animals when diseases like foot and mouth crop up. People get scared and accept it. At times, when the mood of the people is right, we can override these values. When we have the spirit of Londoners during the Blitz."

"Yeah, but then you need someone to hate," said Darren. "Hitler was easy to hate back then. Who's going to be your hate figure now? Yourself? You'd go bonkers."

"Bayern Munich?" suggested Joe.

"That's the other thing," began Doris. "They all say it'll be so easy. They all say, we don't need to change what we do, the way we live. We can go on flying. We can go on driving everywhere. We can have our houses as hot as we want and eat as much of this and that. They daren't say it'll be tough."

"She's right," said Joe. "You have to tell them. Like Churchill. If you tell lies, if you hide the truth, you say it'll be easy … then you'll never get people on board."

"I believe you're absolutely right, my friend," said Igor. "We've made things harder for ourselves. We've created a web of values for good reason – from hard lessons – in order to make a better society. Now we're trapped in this web, and the values make it harder for us to address the problem of greenhouse gas emissions. The only thing which can override this is the mood of the people."

"You mean," said Joe after a while, "if people are scared enough, they'll let the politicians do what they want." Doris shuddered and went to put the kettle on.

18. Disasters

Why it's no use hoping for disasters.

"What about disasters?" I asked after we'd thought about that. "Won't disasters change the mood of the people, and let us take more drastic action?" The Professor had dropped his digestive into his tea and was using his finger to retrieve the bits. "Disasters are very interesting," he said, "but before they work in our favour, they have to meet a number of conditions. An oil spill in the Gulf of Mexico won't persuade us to use the bus.

"First they need to be close to home. It's no use a million

Bangladeshis drowning in floods. It's too far away.

"Then, people have to see that the disaster was caused by the problem in question. It doesn't help us if everyone thinks it was caused by sun spots or bad hygiene or poor safety regulations. Scientists will have to get a lot better before disasters will work.

"Then, there's the problem of forgetting. As the memory fades, we forget the lesson. So you want disasters to be repeated from time to time, lest we forget."

"It's hard to think of something like this. Unless you're an Accrington fan. Their disasters are repeated week in, week out," said Joe.

"And that is not enough either," continued the Professor. "The disasters should be repeated, but they should not be predictable. If they're predictable, then we will start to adapt to the rhythm of disaster rather than fight its cause. And it needs to be severe enough for us to wish to avoid it in the future, but not so bad that we slump into defeatism."

Joe agreed. "Right, look at the crowd sizes at Accrington. They have given up hope."

"I can't think of any series of natural disasters which would meet these conditions," I mused. "Could huge blobs of methane hydrate be released from the sea at unpredictable moments, float over Washington, and go on fire just as they arrive?"

"Don't confuse God with Terry Gilliam," said Igor. "The probability of this is even smaller than that of Burnley winning every game. We can't rely on angry nature to make us see sense."

This dark note left us in a sombre mood. It even put the defeat at Hull into perspective.

19. Other animals

The climate change thing is not just about humans.

Midweek the Professor said to Joe there was someone he wanted us to meet. "Crumbly has taken a strong interested in greenhouse gases in the last few years. He has an interesting perspective," said Igor. "We shall have to take a car to Bashall Eaves and meet him in the Red Pump. They do an excellent hare pie."

We found the General talking over the fence to some cows in the next field.

"Bashington-Smythe," he introduced himself, crushing Joe's hand.

"Pleased to meet you, General," Joe winced.

"Call me Crumbly. They all do," smiled the elderly solider. "He's got you into all this has he?"

"Well, sort of" mumbled Joe. "I didn't think he'd got the military involved yet, though."

"The military? Not at all. Nothing for the military to do, I'm afraid. Can't be taking pot shots at Porsche drivers, can we? No, something far more serious. Lions and tigers."

At lunch the General explained. "When I was in India and Africa, you see, a lot of the chaps spent their time shooting these poor chappies. I never had any time for it myself, none at all. I have never really been able to see why people want to shoot things they won't eat. How can they see themselves as any better than other animals? Humans are always picking fights, showing off, trying to be better than the other ones, and far too clever if you ask me."

He turned to Joe and asked him if he'd ever seen a lion.

"No, not in the zoo, my boy! I mean in the wild. On the plains. Hunting gazelles in the bush. Or a tiger stalking water buffalo. Breathtaking, absolutely breathtaking. The rippling of their muscles, the sheer power, the speed, the utter concentration as they go in for the kill. Highly arousing, I must say." His eyebrows darted like great white moths. Then they fell. "Although Grace never agreed about the lions. Poor, dear Grace. In the end she was eaten by one, you see. My wife. There's nature for you. Powerful, beautiful and cruel, but always utterly dignified.

"An animal never loses its dignity. I've been keeping bees for over thirty years, you know. They keep their heads down, the little beggars. Work, work, work. Never complaining. Then at the end, they just fade away. Ever so quickly. No moaning. They just go."

"Look at those," he indicated to a group of youths at a table on the far side of the courtyard. They were in their late teens and well the worse for wear, loud and unpleasant. "Disgraceful. Noisy creatures, making trouble. They lost their dignity as toddlers when they were fed on Tellytubbies. A culture doomed to fail."

The General fell silent for a while and tucked into his hare. After a few mouthfuls he continued.

"There are no grounds for thinking that humans are in any way superior to other living species. What about the sequoia tree? Could there be anything more noble than that? I'd give a dozen of these louts for one sequoia tree.

"And once you understand that humans and animals and trees are all terribly similar, and when you feel humbled by the speed of a cheetah, the beauty of a peacock, the grace of a gazelle, the serenity of a sloth, the dignity of a dying woodlouse, … when you start to know your place on the planet, then of course your politics start to change.

"Oh dear," he said, wiping an eye. "Silly, silly me. I've gone all blubby. Silly me. Who'd have thought I was once a general." Joe handed him a serviette and he pulled himself together.

"Once you begin to see animals and trees and rivers and other living things as wonderful things on the same level as humans, then you no longer agree that people should be free to do what they want if it destroys the habitat of these poor souls. You believe that we should tread as gently and quietly as possible on the land and on other species. You no longer indulge the fools who build sprawling shopping centres and golf-courses everywhere, quad-bike courses. You would impose drastic penalties on damage to natural habitat, illegal import of endangered species or tropical woods. Oh, I can't list it all now, but you know what I mean.

"Look how they live. People in cities. Squalour. Stress. Inhumanity. Gruesome ambition. Poison, poison, poison. How can that be better than the life of a gazelle?

"All our philosophy and politics and ambition and lifestyle and the whole blasted economy – it's all centred around humans. It's utter balderdash. The biggest tragedy is that hedgehogs can't vote."

Joe pointed out that hedgehogs couldn't really organise a revolution either. People would run them over with big trucks if they started congregating threateningly.

"True, my boy. We assume that animals are under the dominion of men. Even the church has said that for hundreds of years, when the church should be promoting the preservation of God's earth."

The Professor had remained unusually quiet. I think it was the food because as soon as he had wiped his plate clean – with his usual vigour – he spoke up: "And what about the hare pie, then?"

"You can't catch me out that easily, Igor. I eat meat once a week, no more, no less. I usually shoot the beggars myself, skin them and gut them. It's fair game."

He turned back to Joe. "What can be done, my boy, is to start 'em early. Start 'em very early. Instil in them at the age of three, four, five … a deep, passionate and indelible respect for nature. If it's in them then, then it should stick. That's the trick, my boy. It must be visceral, not intellectual. An automatic response. An involuntary sense of revulsion towards cruelty, dominance of other species, wanton destruction, wilful causing of pain. And an instinctive sense of balance.

"With this way of thinking, people would be hugely more receptive to your politics, Igor. Tough laws on emissions might seem authoritarian today. But to people properly raised in the spirit of communion with animals, they seem simple common sense.

"You might be able to change adults. I don't know. There are too many socialised in the cities. Their sensitivity to nature is crushed in a landscape of concrete and steel and glass, by constant noise and grating. The cure would be long and slow and costly. There may have to be significant disruption to their lives."

He sat back in his chair. The moths settled and he appeared to doze off for a few minutes. "But if you're ever in London, don't for goodness sake mention all this to the chaps in politics. It'd go over their heads. Now, thank you for the delicious lunch, but I must be going. I have my bees to see to and not so many years left as you."

We pulled out of the carpark and immediately Joe let off steam. "What the hell was he on about? All that about lions and shite and the crying?"

"An unusual chap," chuckled Igor. "He's saying that people will always harm the planet as long as they don't care about the other things which live on the planet. And to care about things, you have to get close to them. Would you worry if Torquay United went under? Or Grimsby or Lincoln or Aldershot or Plymouth or QPR? Or Watford? Or Tottenham and Chelsea and Portsmouth?"

"Course not, it'd be great," said Joe.

"Until the football league falls apart. And then you'd have no-one to play against. The TV revenues would dry up, and that'd be the end of Burnley, too."

"Right, but that doesn't mean I want to get all touchy feely with Chelsea fans. I mean – " Then he remembered abruptly that Chelsea fans were a very sore point with me.

"Not touchy feely," said Igor. "Just respect, that's all."

20. Vested interests

Many people and organisations are set against cutting emissions.

It was bonfire night and time for Beşiktaş to wreak their revenge. Mr Trew said that a week off for one match was taking the mickey and the Professor refused to fly. Uncle Frank went, but Joe and the rest of us stayed in Burnley. It was cold and wet and November-ish. So we sat at Joe's and tried to keep quiet because Darren was in trouble at school and had to study. We were too nervous to be noisy anyway.

There were over 31,000 Turkish fans and a few hundred

Burnley fans in the ground.

"Trouble is," said Joe to the Prof, "all your ideas just won't work. There's just too many folk on the other side. How're you going to persuade them all?"

"What do you mean?" asked the Professor as he looked up from his notes. He rubbed his beard, sensing debate.

"Well, first you've got people running power companies and oil companies. Like that lad, Pete at number 38, made it big in oil. Always flying to China and Saudi Arabia and Washington. He's not going to suddenly start rocking the boat, is he? He'll be out of a job. You'll not shift that lot."

The Professor said that there are only a few of them. It's lonely at the top.

"You might be right," countered Joe. "But we're not talking about a street fight, are we? In any case, they'll not back off without a struggle."

"It's goodies and baddies, no?"

"They might be baddies, Prof. But they're you and me, too. Anyway, then you've got the folk who are too clever by half. Lawyers, solicitors, politicians, professors, that lot. It's far more important for them to win an argument for the sake of it than to help some bleeding budgerigar in Bangladesh. Whether they're right or wrong don't matter.

"Then you've got the economists. All they want is to make the economy bigger. And they've us all thinking that getting bigger and more and better is the only way."

"Never happy, that lot." Doris shook her head. "Much wants more."

"So that's it, Joe? You have the oil people, the too-clever-by-half brigade, the economists … oh and the press, of course?" asked Igor.

"No I haven't finished yet." Joe snapped open another can of Stella and passed one over to the Professor. "Fancy another?"

Doris laughed that at least there was one bit of football culture that the Professor had picked up.

Joe added to his list the people who don't give a sod for trees and birds and animals. Millions of them. Folk who were brought up in cities think cows grow in freezer chests. Then you've folk who have to prove themselves – money and ambition. They can't handle people never mind cats and dogs. Never mind coral reef thingies. Then there's folk that reckon there's no problem because you can always build some big machine to solve it. And then there's just folk that don't care a monkey's about anything."

The teams lined up and the UEFA anthem was played.

"So," said Igor. "The opposition crowd includes owners and agents of heavy industries and their financiers. We have politicians, lawyers, and academic people – the intelligentsia who use their intelligence in the pursuit of their own glory and not in the pursuit of truth, justice, and natural beauty. Then you have the "don't-care-a-sod" people who by way of socialisation are inimical to environmental concerns and environmentalists. And finally all the don't-give-a-monkey's people."

"Like I said, it's a big crowd," said Joe.

"It means we have to shout all the more loudly."

The Burnley contingency was vocal. It became even more vocal in the fifteenth minute when a through ball by Cork was picked up on the edge of the area and without a blink Eagles curled it just wide of Rustu's far post. It was a warning from the men in claret that they meant business. Only minutes later a short corner was flicked on by Carlisle, and Paterson at the back post, finding himself unmarked, couldn't miss from three yards. Feverishly as they fought, the Turks couldn't muster a reply. The one goal was enough. Job done. Delirium.

21. The thin blue line

A reminder that markets are not good at solving some problems.

The next home game wasn't such a success. It was a half past one start and we were beaten soundly by Aston Villa. The 2-1 score-line flattered our game. We remained on ten points with December imminent.

As we left the ground Igor motioned us to stop. "Just a moment," he said, looking back towards the ground. We stood and watched the crowds streaming out of the ground onto Harry Potts Way. Everywhere was cast in late November drizzle. The sky was darkening, heavily overcast, with just a gleam of winter sun over the rooftops. The reflections of traffic lights, cars, and front-room windows glowed in the damp. The crowds edged forward through the rain, sometimes purple and blue umbrellas dancing over their heads. Infantry tramping back from battle on the moors.

Police cars stood by, their blue and red lamps sweeping across the faces of the supporters. Policemen directed the flood of fans out of the ground and up Brunshaw Road until they gradually dispersed into the dark.

"Order," said Igor quietly.

"You what?" said Joe.

"You see how orderly they all are … the crowd filing out … thousands of people shunting slowly out of the ground and away. No disorder, no panic."

"Well that's what the police are there for, isn't it?" said Joe.

"The masses … they … we need the forces of law and order. We need authority to guide us. There's no economic market which would ensure the orderly exit of 22,000 fans from Turf Moor."

"Yeah," said Frank. "Without the police it would be bloody chaos."

22. A brush with the Carabinieri

How long do you go on accepting the will of the majority even if you can see that they are killing us all?

We were on the way to the Lazio game. If sitting on a train for 24 hours doesn't turn you into a philosopher, nothing will. So to be safe Doris brought plenty of Sudokus and an armful of celeb mags, and Joe packed away a dozen cans of beer. Meaning I was left to Igor and his thoughts.

He was talking about the relationship between justice, violence and the generals. The guarantors of justice and stability aren't our politicians but men with batons or guns. It's easy for us to forget this in peaceful England. You only notice it when there are football riots or protests about petrol prices. But in Russia or Turkey or Chile you see the military on the surface of society.

When people disagree in our society they debate and argue. Eventually one side gives up and walks away. But it's different in other places. In countries like India or Pakistan or Iraq or Kosovo arguments quickly turn to fighting. They start having a squabble about who forgot to pull the chain, and in no time they've got knives out and are chopping each other up.

So if we feel very strongly that we have to cut emissions but we can't persuade everyone else to change their ways, what do we do? When do we say: 'I can't win this argument with words'? We go on watching emissions rising, watching politicians around the world being ineffective and prevaricating and hesitating and being disingenuous and incompetent and corrupt, we go on watching vested interests sustain the status quo, and we can't persuade them with words. When do we say: 'Time's up, this is something we need to fight about'?

In our society if you start a fight, the police or the army will roll out and mop you up. This is because they have a deal with the government. They get the best kit and lots of kudos and good pensions and get to drive tanks and shoot machine guns. In return they have to bat on the government's side.

Stability is maintained because everyone knows the generals stick up for the government if there's serious trouble.

But the government has to be sure that the generals will come down on their side when push comes to shove. So the actual beliefs and opinions of the generals are critical – the politicians know they can't do radical things which the generals strongly disagree with.

If the generals and the government are in disagreement about something, the government and the country are at risk, since the commitment of their protector is weakened.

Therefore we want the generals to be deeply convinced that big cuts in emissions are necessary. If the government knows that the generals believe this, then the government can be confident with policies to make big cuts. And if your army is big enough and has international reach, then it also helps other governments with their policies.

Already senior figures in the US army have called for action on emissions. Their views will influence governments which are lagging behind.

"You see," said Igor. "It's critical to get the generals involved. You never know when you'll need them. You can't be sure that everything will go very smoothly."

It didn't. Outside the Stadio Olimpico, Frank swore at a Lazio fan who got in the way while Frank was taking a photo. So the Lazio fan knocked Frank's camera to the ground. Frank knocked the Lazio fan to the ground. A dozen Lazio fans knocked Frank to the ground. Soon we had a scrum and then a phalanx of Carabinieri swarming across the Piazzale to mop up the riot. Frank was carried away struggling like a bullock, and the Lazio fans scarpered, waving Frank's scarf as a trophy.

23. When in Rome

Then there is the small matter of the rest of the world.

The match itself wasn't too bad – a 1-1 draw with Alexander scoring an 89th minute equaliser from the spot. We were still hanging on in there.

"Horrible place, Rome," moaned Doris when we were settled on the train back. Minus Frank, who was coming home under separate cover. "The coffee was awful, don't know how they can drink it. The pasta was chewy. The ice-cream was too cold. The traffic was a nightmare. And they don't speak English properly. And they've taken our Frank. Why would anyone go there if it weren't for an away game?"

Igor sighed. "When in Rome, do as the Romans do. What you consider undrinkable coffee is for a Roman perfection in a small cup. The pasta you eat at home is for an Italian soggy gruel. When traffic is made up of Alfa Romeos – who cares how long the jam is? Each country has different views and different customs."

"No excuse for cheating, though. Their goal were miles offside," remarked Joe.

"It was. But look here. All the Italian press here says that Eagles dived and we shouldn't have had the penalty. Many things depend on where you see them from."

Igor was quiet for a while. "Come to think about it, these

differences between nations and peoples present a challenge for those who wish to cut greenhouse gas emissions. All these policies might work in England or northern Europe. And Costa Rica. And a few islands which will shortly disappear. But that's it. There are seven billion people in the world, and only a few hundred million are in countries where the political system has the genuine will and capability to do something about it.

"The EU wants to be the world's centre-forward in climate policy. The one scoring the goals. At the moment it can only score own-goals. It can't even get its own emissions policies to work – and that's between allies, neighbours, friends and partners with hundreds of years of common culture. What's the chance of getting a deal between 190 countries which are enemies or rivals or have centuries of animosity between them?

"So it's no use people in Britain thinking they can reduce their emissions then everything will be fine. Emissions have to be reduced in all rich countries and many poor countries. And with a few exceptions, they will have a much harder job than in England."

"Why's that?"

"Their people have no interest in the problem. Or they aren't used to thinking long term. Or they're less interested in conservation. Or they're ill-educated. Or the problem's very low priority for them. Or the countries are politically dependent on oligarchs with no interest in the future. Or it's culturally acceptable to have the heating on full blast and the windows wide open. Or the countries are too conservative to change things quickly. Or they're cluttered with bureaucracy, stifled by corruption. Or all the energy engineers have left the country and there's no-one who can build wind farms, nuclear power plants, or clean coal facilities. Or they're bankrupt and can't raise finance for such changes. Or they're more interested in fighting wars with their neighbours or ethnic minorities. Or they're already deep in it – crying for help and no-one listens to them."

"Perhaps they're just more worried about where their next meal's coming from," suggested Doris.

"So what you're saying," said Joe. "… it's a complete f– "

"Joe," warned Doris sternly.

"In different words," said the Professor.

"Well what do you do about it?" asked Joe. "What's the point of us worrying about all this if the rest of the world doesn't give a toss?"

"A very good question, my friend. In fact, we only need the USA and China to take it seriously. But even they have other priorities."

"You know what you could do," said Joe. "We spend billions on weapons each year. We could use them on the countries which can't get their act together. It would soon sort them out. We've got the best army in the world."

He thought for a moment. "Or will do, once we get the helicopters fixed."

24. When in Bonn

International efforts to cut emissions are dismal. It's the way they do it.

Somewhere outside Milan Joe fell very quiet. He was working on an idea. Then he told us about it. "You see there's this organisation in France in charge of international football. Called FIFA. Or it might be Switzerland. Say they want to change something in football, like change the offside rule or arrange the World Cup or something, then that's what FIFA does. So I thought, why don't you have an international organisation to do something about your emissions? I mean with football they all have the same rules to play by, and it goes for everyone. Same rules in the UK, Togo or Taiwan. Why don't you have an organisation which does this for emissions thingies?" said Joe. He was still wide awake with the expresso from Rome.

The Professor said that there is already one. "It's called the United Nations Framework Convention on Climate Change. It's based in Bonn in Germany. And it's part of the problem." The Professor pulled a sheet of paper from his notebook and waved it at us.

"This is a transcript from some recent international climate negotiations in Germany. This is how countries negotiate with each other to try to agree to cut emissions."

He read out to us...

Representative for Kenya, Chairman of the meeting: "Welcome to the continuation of the fourteenth subcommittee meeting B ... er ... I beg your pardon ... subcommittee meeting (b) of the committee on modalities for reporting uncertainty in estimates of the impact of air-conditioning regulations in the Inuit territories of Nunavut, Nunavik and Nunatsiavut.

"Yesterday, we adjourned at paragraph I.1.a.(i).A. Australia?"

"I question the use of the word 'the'."

"Which use of the word 'the' is that?"

"The third instance in your previous sentence."

"You mean 'the impact'."

"I do."

"And what is your proposal?"

"There's a logical inconsistency. If we are talking about the uncertainty of estimates, then we can't be sure whether there's any impact of these regulations or not. Therefore, we propose to replace 'the impact' with 'impacts'."

"Objection."

"Poland?"

"First, may I thank you for giving me the time to address this august body. I recall some twenty five years ago when I first met Mr El Shawabi at an artificial snow trade fair in Abu Dhabi, what a deep impression his sincerity and wisdom left on me. I hope that we can continue to benefit from that during this session. Second, I wish to remind all those here that today is the birthday of the great Polish scientist Zygmunt Wroblewski. He reported the discovery of the carbon dioxide hydrate while studying carbonic acid, a major step forward in mankind's understanding of gases. I hope that we can all draw inspiration from this great man. Third, I suggest that we avoid the term 'impacts' since that implies that there's more than one impact. This could lead to an unnecessary search for impacts, thereby wasting critically important time and resources in the search for policies and technologies, whatever they may be, and whenever they may be uncovered, discovered, or even chanced upon. Therefore our suggestion is to replace 'the impact' with 'the impact or impacts'. This avoids the possibility of assuming, rightly or wrongly, that there are many impacts, while opening the door for the possibility thereof."

"But, Professor Jaworov, in the English language the word 'impact' actually has a sense of both the singular and the plural in its singular form."

"Gentlemen," spluttered Bolivia. "This is outrageous. I object in the strongest terms to the patronising style of Australia. Just because his mother tongue is English, he should not lecture to the Polish gentleman about English semantics."

China wished to be seen as the dealmaker, brokering a solution: "We propose to replace 'the impact' with 'the impact or impacts, if any'".

The USA responded: "We recommend replacing 'if any' with 'to the extent that there might be'.

Australia: "We can live with this if you insert 'proven to be' after 'be'.

"Gentlemen, time is moving on. I am asking for compromise."

USA shook its head: "We can't move on this."

Australia: "May I request a five minute adjournment?"

"Agreed. We adjourn the meeting and continue at 2.55pm."

The Professor folded up the sheet of paper and continued.

"This is a record of just one meeting. Hundreds and hundreds of meetings happen like this during the UN's climate summits, held at exotic addresses around the world," he said.

"Most of the meetings are a waste of time and of taxpayer's money. They're run by bureaucrats and professional negotiators. The best are paralysed by the complexity of it; the rest aren't up to the job, shoved into positions because there was an empty box which needed a name. They forget why they're there. As negotiators they're trained to win negotiations, not to help the planet."

Doris said that it means they couldn't see the wood for the trees and we had a chortle.

Igor folded back the script from Bonn and explained. "These negotiations are a hangover from the idealism of the United Nations. It was obvious several years ago that this approach was doomed to fail. Yet for some reason they persisted with it doggedly, in Bali, Poznan, Bonn, Washington, and Barcelona. In the end the monstrous bureaucracy of the UN won over. Remember the fiasco of Copenhagen in 2009? 200 million dollars were spent on organising the summit, the culmination of two years of preparations. And on the day the Heads of State arrived, there was nothing for them to negotiate. They were humiliated. The UN had completely failed. Any organisation which gives Hugo Chavez an international podium for half an hour ..."

Then he continued. "Even if they got something on paper, any agreement binding 190 nations and seven billion people is bound to be so weak that it's useless. The only agreement they can achieve is not enough. This is the limit of international politics, and it's not enough."

Joe said it would be like trying to play the Premier League on one massive pitch with all the teams playing each other at the same time. Igor replied that it was irrelevant anyway. Over 70% of the world's emissions are caused by six countries or regions – USA, China, the EU, India, Japan and Russia. Why make the process impossible by involving Togo and Laos? It would be like negotiating the Premier League schedule and involving in the discussions the Championship, League One, League Two, Blue Square Premier, Blue Square North, Blue Square South, and Droylesden Firemen junior five-a-side league.

Then Joe put an end to the discussion: "You know, Prof, if something's really important the only way you can win is if you get tough – it's no use just talking. You either put sanctions on them or you do regime change. If they're not ready to go all the way, it just means it's not important enough for you. Mind, if we're going to do it, we'd better do it while we're top of the pile." We were back to the helicopters.

25. The time it takes

It takes a long time for ideas to become mainstream, especially when they have to be proven scientifically. But we don't have much time.

"What do they spend all their time doing, these people?" blustered Darren. It was a cold and windy Saturday morning, the first in December. We'd gone down to Portsmouth a day early. Joe and Frank stopped in the pub and Igor suggested Darren and I take a brisk walk with him on the quays. The sky was clear and we could do with some fresh air.

Darren wanted to know something. "They've known about this problem for ages. Why does it take them so long?"

"It's true, scientists have made heavy weather of climate change." Igor gave us some history. It had taken over 180 years since a Frenchman – yes, a Frenchman! – realised that the heat from the planet would be trapped by certain gases. That was in the 1820s. Now after 180 years we've got to the point where many scientists finally agree that we are, as you'd say, up the creek. Most scientists believe that the creek is the one on the map called "Shit Creek".

"Right, and most reckon we are there without a paddle," I said.

"True, although there are still some looking in the water for the paddle while we approach the rapids. Just think – it took scientists 180 years, and yet they have university education, PhDs, professorships, and huge laboratories for battling with the complexity of the carbon cycle and the oceans and El Niño and La Niña and sun spots; they built mammoth computer models to predict fifty or a hundred years hence. 180 years to get to the first post. How can we expect the general public to be any faster at accepting the problem?

"But climate change is nothing special, Darren," continued the Professor eyeing a seagull eyeing his sandwich. "Think of smoking. They knew smoking was harmful in the 1600s. We only had scientific proof it causes cancer in the 1960s. And even now despite scientific concensus, swingeing taxes, public education, active discrimination and stigmatisation, still 25% of adults smoke in England. And I don't blame them. It can be very satisfying to belong to a threatened minority.

"Or television. With a bit of common sense you could see the harmful effect watching television had on children, as long ago as the 1970s. It made them violent and anxious and gave them difficulties concentrating. Yet only now, thirty or forty years later, are scientists starting to demonstrate these connections. Legislation limiting television watching is decades away.

"Science is slow, Darren. Common sense is far swifter. But common sense is not always reliable, it can be distorted by fashion and convention."

Darren still didn't understand why it takes them so long to get their act together. If they were at Dad's work, they'd've been kicked out ages ago.

"Science is very complex. Things must be proven step by step. Each step is a whole career. Knowledge is the compression of physical matter into intellectual form. Just because knowledge takes up little space, don't forget that the act of compression requires the effort of many men and women for many years."

"So you mean while the planet is burning, scientists are still trying to understand why it's burning instead of doing something about it," said Darren.

"You could say that, Darren. But that's not fair on scientists. They're doing their jobs, and working with dedication and integrity."

"Well, we've heard that one before, haven't we, Professor? I mean, only doing their jobs."

"You're harsh, my friend. The reputation of science as the source of truth depends on the reputation of scientists as being objective. And today, the only source of truth is science. Truth grows very slowly, I'm afraid. Like a good wine. Or a tree."

"Then if we want to cut emissions in time, we haven't got time to wait for truth, have we?" concluded Darren.

Igor threw away the rest of his sandwich to the gulls. "We've no time at all. The time when we could wait for scientific certainty is long gone. We have to make a bet."

26. Overwhelming complexity

Climate change is not a normal problem with a simple solution.
It is too complex for that.

"That's part of the beauty of football, Prof. It's completely illogical. We lost 2-1 to Villa the other week. Portsmouth beat Villa 2-0. And now we beat Portsmouth 3-2. Away! But that's football. It's crazy and beautiful," Joe was saying.

"I understand that it's a complex system with unpredictable elements. But why is that beautiful? A woman is beautiful, yes, I agree. A view of the Pennines is beautiful. But football? No, it's absurd," argued Igor.

"No more absurd than your climate stuff, Prof. That's way more complicated than football, and it just makes us depressed. At least football cheers us up."

"If you're on the winning side," Igor reminded Joe. He looked out of the coach window. They were still miles from home, the motorway traffic around Birmingham was an endless tangle of lights with no end and no beginning, blurred by the rain streaming across the glass.

"Right. But how are you going to get on to the winning side with your carbon emissions? You're not. It's just too complicated," said Joe.

"But ..."

"I don't mean the problem itself. That's easy. It's getting hotter and that's bad news. We know it's trouble. And we know what causes it. You just have to join up the dots. You're sitting at home, switch on Match of the Day, into the plug, down the wires, through the power station, up through the chimney and out into the sky. That's easy, isn't it?"

Joe continued. "But it's doing something about it, that's the complicated bit. There's some stuff which you can just solve easily isn't there? The lights go out, so you check the fuse box. The cat's not eating, so you take her to the vet. Darren's forgot his mobile so … you know what I mean?

"And then there's problems which are so complicated and big that you just panic and your head goes mental. Like you find grandpa in his bed and he's not breathing. Like if you can't pay the mortgage and the electricity is going to be cut off and Cathy's off work with her back and we can't afford a birthday present for Kev. Them kind of problems … you just … I mean, what do you do?

"Your carbon dioxide is like that. You tell us that driving to work is part of the problem, and holidays abroad are, and eating meat is, and going over to the east coast to visit granddad is, and we've the wrong kind of windows and walls and roof and door and heaters and … and shower and cooker and washing machine and dishwasher and coffee machine and lawnmower and light-bulbs…. And we've got to live off three bloody Smarties a day. When you think about all that … you think … bloody hell we're all dead in the end anyway."

Igor was quiet for a long while. "Joe, you are absolutely right. Climate change isn't a problem like normal problems. It's more tangled than the Spaghetti Junction. I don't think it's a problem which has a solution – there's not a white knight. Some great global agreement isn't going to make it go away. But what we can do is do our best – our very best – to tame climate change. And if we do it right, we might be able to help ourselves in lots of other ways at the same time.

27. Real progress

Where we've got to.

PSV were to visit Turf Moor the coming Thursday for the last game of the Group Stage. The Christmas lights had gone up in town, the shop windows gave that tingly feeling that Christmas is around the corner. Anticipation and celebration were in the air.

Joe and Doris had us round for mince pies before the PSV game.

"I baked them myself," said Doris to Igor's delight. Joe poured him a sherry.

"So Prof. We've been at this since summer. Are we making any progress?" asked Joe.

"If we win tonight, we will have. I understand that it'll take us through to the so-called knock-out stages."

"No, he means with your carbon stuff," said Doris. "You've shown us all sorts of reasons why we have greenhouse gas emissions. You've taken us through the psychological and – what do you call it – sociological reasons; reasons to do with our brains and evolution and our culture and stuff."

"And why it's so hard to change things. Because of the way society's structured. And because of people with vested interests," added Joe. "And the values we have. And all the different cultures and countries. And the way international agreements are made. It's a complete –"

"It's a complete pile of bollocks," said Frank.

"So what's going to be done about it?"

The Professor picked at the crumbs on his trousers. "Before I can answer that, we need to take a different tack. We first need to look at the different things which governments are trying to do at the moment. Let's see whether the things in place are ever likely to work."

"Fair enough," said Joe. "But we can have a rest over Christmas can't we?"

"Of course. Let's hope the Clarets give us the present we want tonight."

God was wearing Christmas-edition claret and blue pyjamas that night. For a couple of hours Turf Moor became a cathedral, and the crowd its congregation. The loudest carols ever sung, with the greatest passion ... they must have been enough to wake Him. For Eagles seemed to sprout wings as he hurtled down the left wing after a long ball from the middle. He floated past Ooijer at right-back, graced the by-line and his sublime cross found the flaming head of Thomson with exquisite precision.

We were through to the knock-out stages of the Europa League!

Part II
December to February

Governments' efforts to cut emissions are even more disappointing than the Clarets' mid-season.

28. Christmas crisis

A crisis came at Christmas. We had four games. After a draw at Wolverhampton we picked up no points from the next three matches. We were trounced 4-1 by Arsenal and beaten 3-2 by Bolton then lost 2-0 at Everton. It was a chilly and frustrating end to the year. Christmas was very lonely at the bottom of the league. Even Hull picked up more points than we did.

"It's a crisis," said Joe.

"Of astronomical proportions," confirmed Frank.

"I just don't know where we start."

Igor said: "I see the Burnley board are making noises that we need change. Same with my little climate crisis. Lots of noises, no real action."

"Meaning, we're stuffed," said Frank. "Talking of which, any more of them turkey sandwiches, Doris?"

"And the politicians think we can carry on pretty much as usual. Once you've said that, there's not much you can do. For that matter, most people think we can carry on pretty much as usual."

"Like, you're bottom of the league … and you have to win at least eight more games this season to survive … and the Gaffer says you don't have to change the way you play. That doesn't sound right, does it?" said Joe.

"Well, they're trying to do something. There's a feeble effort at making an international agreement to cut emissions by a certain percentage by 2050 and by 2020. Unfortunately this is going nowhere. Then they have big plans to bury carbon dioxide underground. And they're introducing schemes to put a limit on the emissions which some industries can make. This results in putting a price on carbon emissions. Then … well there are bits and pieces – schemes to replace boilers and promote insulation, rules for building new houses, programmes to build wind farms."

Igor looked at the half-eaten piece of Christmas cake on his plate. It had been sitting there some time. "In fact you might find the list of what governments are doing – and why it's all messed up … you might find this all fairly depressing. Think of a season where you just go on losing week in week out."

"Er, like this season, you mean..." said Joe glumly.

"Far worse. It's quite pathetic, really. They're …"

"It's like …" began Joe.

"Rearranging deckchairs on the Titanic," said Igor.

"It's like," offered Uncle Frank, "trying to escape a pit-bull by tickling its belly."

Doris came in. "We've run out of Hobnobs. Will Lincolns do?"

"It's like having a major global biscuit crisis, and suggesting we make do with Lincolns," said Joe.

"Inadequate," tutted Darren. "Wholly inadequate."

29. The wrong targets

Some problems with targets to reduce emissions by a certain percentage.

"So what's this 80% and 90% about?" asked Joe. We were stomping about on the terraces trying to keep warm, waiting for the start of the Newcastle game. It was the first game of the new year.

Doris poured out the thermos. "Bit of global warming for you, Professor."

The Professor had a bad cold but managed a brave smile.

"To get worldwide emissions cut by 50% by 2050, the rich world has to cut emissions by 80% or 90%. This would give us a 50% chance of avoiding warming of 2°C or more." He took a coin from his pocket and threw it up. "This much chance. Would you bet your house on this?"

"The bit I don't get, Professor," began Joe, "is how come it happens that … it's exactly fifty percent by 2050 … I mean, that's a very handy numbers."

"It's a soundbite," explained Igor. "People need soundbites; headline writers need very short sentences. 'A 50% cut by 2050' is good because it's five words with poetic repetition of the word fifty. '20% cut by 2020' has the same feel. But these round numbers have a slippery deception. These are quite the wrong targets. We should legislate for larger newpapers to accommodate longer headlines. You see, the limited bandwidth of journalists is driving the science."

"But wouldn't bigger newspapers mean more paper being wasted?" asked Joe.

"He's joking, you prick," said Uncle Frank. "But if it's only about 2050 then I'm buggered if I care. We'll all be long gone by then, so will the politicians. If it were that important it wouldn't be about 2050, would it? It would be about now, 2010 and 2011."

The Professor agreed. "A few numbers can't be a political commitment. It has no meaning. As long as you don't define it, politicians can just skirt around the truth, making it all sound easy."

"So what does a 90% reduction actually mean, Prof?" asked Joe.

Igor looked to the sky, formulating his words. "It means very big things. An 90% reduction in emissions can't happen with little changes. It won't happen by snipping nervously here and there like a trainee hairdresser, trimming little bits off the edge and keeping all the big chunks that each vested interest wants to keep. Cutting emissions by 90% means practically eliminating all emissions except for special cases. Everything that we do will be affected very significantly.

"Most of the houses in England will have to be knocked down and rebuilt – patching them up's probably not be enough. Petrol cars will have to go. Flying to meetings in New York or skiing holidays in Val d'Isère will have to go. Cement and steel and glass might be too expensive to produce in a low-carbon economy. Meat will become a once-a-week treat on Sundays."

"Bloody hell," Frank whistled. "Think of all them companies that'll go bust. Shell, BP, Exxon, Chevron, BA, Heidelberger, Holcim and Arcelor. Bloody hundreds of them. That'll be a laugh."

"So it's not really about numbers at all," said Joe.

"Of course not!" exclaimed the Professor. "It's about real things. Real lives."

Then Igor got out his notebook. He said there's another problem with setting distant targets. It leaves a lot of wiggle room. He showed us two graphs.

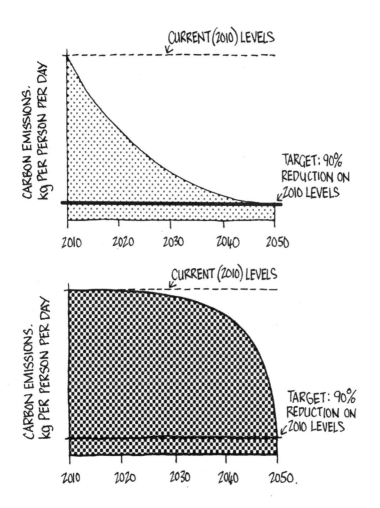

"Both these graphs have a reduction in emissions of 90% between 1990 and 2050 But in the bottom graph the total emissions are 40 billion tons of CO_2 and in the top graph the total emissions are just 15 billion tons.

"The first graph results in a catastrophic outcome and the second one is possibly safe. Long-term targets are just avoiding the real challenge."

"Flipping heck," said Joe. "So reducing emissions by 80% by 2050 could mean anything."

"Absolutely. Even when they talk about 30% reductions by 2020 there's enough leeway to make a critical difference."

Frank said this proved that it was all a pile of bollocks. Before the Professor could respond, the players came out so we avoided a show-down.

The Magpies played a tough game. But the Clarets were up at half time thanks to an inspired free-kick by Alexander.

"Brilliant player. Old enough to be a grandfather, mind," said Joe.

"Grandfather ..." said Igor. "That reminds me of another problem with these targets."

"They're all old bastards, right?" said Frank. "The people setting them. They'll be dead, won't they?"

"That's true, but not what I meant. I was thinking of something called grandfathering. This is when the amount of pollution you're allowed to make is based on how much you polluted in the past. Grandfathering means that the more you polluted in the past, the more you're allowed to pollute in the future."

"Come again," I said. "How's that work? It doesn't seem to make sense. Shouldn't it be if you polluted more in the past you should be allowed to pollute less in the future? Like to make up."

The Professor explained that under grandfathering, countries with high emissions have the right to continue emitting a lot, and the countries with small emissions get stuck with their current low level.

"But ..." began Joe.

"It's true," insisted the Professor. "On the face of it the EU looks like a good boy – it says it will cut its emissions by 80% between 1990 and 2050. But look where it's starting – at about ten tons a head each year. In contrast the people of Chad emit about 400 kilograms per head per year.

"Fair enough," said Frank. "Look, if Chad kicks up a fuss we can bomb the bastards."

"Indeed, my friend. This is a logic made possible only by the economic and military might of the west."

"Seems a bugger that," said Joe. "Wouldn't it be fairer if everyone could emit the same amount per head?"

"You'd think so. If you believe that people are born with equal dignity and value, irrespective of whether in Burnley or Bongor."

"Bongor?" asked Joe.

"It's somewhere in Chad."

"So you're saying that under the international agreements rich countries can emit 10 tons per head and poor ones only emit less than 1 ton per head each year?"

"Roughly," replied Igor.

"Shouldn't the rich countries pay for that? I mean otherwise it'd be like stealing, wouldn't it?" said Joe.

"Countries try to make up for grandfathering by agreeing to make payments from the rich world to the poor world. The rich world is offering around 30 billion dollars a year. But I made some calculations …"

The Professor showed us a table in his note book.

"If we were to pay a fair price for the emissions we make over what we should, the UK would pay €8 billion a year to the poor world. And the EU, €70 billion."

"Just come again," said Joe.

"What he's saying, Joe," I said, "is that if we were being fair about it, every one of the rich countries would pay a pile of money to the countries in the third world. As it is, they're getting a massive freebie."

"Nice one if you can get it, eh."

"It smells like bullshit to me," said Frank. "If you want to cut emissions, it's not bleeding Bongor where you need to spend the money. You need to spend the money in Manchester and Chicago and all across the rich world. That's where the emissions are. It's a crap idea sending money to the countries without emissions when the whole thing is about spending money to cut emissions. They shouldn't send anything to the third bloody world, they should invest it at home where it's needed."

The players were coming back onto the pitch. We were still one-nil up in the eighty-eighth minute when Carlisle tripped up Alan Smith on the edge of the box. The referee pointed to the spot. Ameobi stepped up to take Newcastle the penalty which would keep us stuck at the bottom of the league.

"You see," said Joe in the Bridge later on. "Sometimes it's good if people miss their targets."

We were settled in our favourite nook and he'd bought us a second round to celebrate the first win of the new year.

"But if you're right, Prof," continued Joe, "and these long-term goals are a load of rubbish, why do they try and set them?"

"Hmm…" began Igor. "It's difficult to understand their thinking. But I believe they're there for people who are making long-term planning decisions. People who are building and financing power stations, factories, cement plants, roads, and so forth."

"How does it help them?"

"It's argued that if they know what the price of carbon will be in the long term, then they can plan better."

"Knowing the price of carbon in 2050? That's total rubbish," spluttered Frank. He wiped his pullover. "I don't know the price of bricks next year, so how could anyone know the price of carbon in 2050. It's daft. You can't even guess it."

"Do you really think that people building power plants make decisions scientifically?" he asked. "When were you born, Igor? Decisions about building power plants … they're made between friends based on who's scratching who's back, who likes what restaurant, who fancies who. It's just plain naïve to think it could be otherwise. People in the power business aren't there for a thirty year contribution to human wellbeing. They're there because … because they want to build something where there was just grass and nettles and bushes and crap. They like the big cranes swinging across the sky. They get a turn-on from negotiating a billion dollar deal. They feel macho staying up all night writing contracts. They do it for the share price. For the way the building goes up a little bit every day. They do it because they love the flash of a welding torch. They do it for kick backs. For kudos. For the guy who brings the coal deal. For a house in the country.

"If you want to build a power plant, you make the numbers work. If you don't want to build one, you'll make the numbers not work. That's what you have consultants for. Give the board a good lunch and they'll swallow what you tell them. Come on, Igor. That's how the real world works. 2050 is irrelevant for these decisions. 2020 is just on the edge. 2012 might make a difference."

"I thought for a moment, my friend, that you were taking up poetry. It's a very fine beer, this Hydes. Another round, everyone?"

We emptied our glasses and the Professor shuffled off towards the bar.

"You're right," he continued once he was back. "No-one could predict with honesty the price of electricity or steel in 2020. Could a prudent investor truly believe such a prediction? These decisions aren't based on precise calculations. It's about feelings and emotions, charming and bullying, and the chemistry between people.

"If they really want to give messages to investors, they could be much simpler." The Professor put his glass down wearily. "Mr Mittal. Steel production in its current form will either be unviable or illegal by 2020. Mr Merckle of Heidelberger – cement production will be unviable or illegal by 2020. But you're clever men. Without doubt. Figure something out. Show us how clever you really are."

The bell rang for last orders.

"Come on you boring lot," cried Doris. "It's last orders! Can't you buy a drink for a lady on her birthday?"

30. Marbles

The idea of cumulative emissions rather than targets.

"If these targets aren't right, what should they be saying?" Joe asked Igor the next day. It was Doris's birthday and we were in Joe's living room. Luckily I had remembered to remind Joe, and he managed to find a large glass vase, filled it with water and stuck a rose in it. It stood on the coffee table.

Igor was playing with some marbles in his hand. He took aim and threw one of the marbles, so it landed in the glass vase, sinking to the bottom.

"Good throw, that" said Darren. "You should be following cricket, not football, Professor."

He threw another marble. It plopped into the vase again.

"Have you noticed, my boy ... As you put more pebbles into the jar, the level of the water goes up. If you take one out, the level of the water goes down.

"It's the same with greenhouse gases. As we put more carbon dioxide into the atmosphere the surface temperature of the earth goes up. If we suck it out again, the surface temperature of the earth goes down ... well, after a few hundred years or so. It's that simple.

"Of course, it's not simple at all, it's highly complicated. But for the purpose of our everyday living and therefore policy making, the marbles in the vase are an accurate parallel."

Doris turned down the telly so she could hear.

"Driving to work is like putting a speck of sand into the vase. A billion people driving to work is like a chunk of sand. Running a large coal-fired power station is like plonking a pebble in. Every bit, big or small, makes the water inch up the side of the jar. That is, it notches up the temperature a little bit.

"In fact, scientists have figured out that between 2000 and 2050 we can afford to put about 750 billion tons of CO_2 into the sky and no more. We already put in 300 odd billion tons by 2009. So we've another 450 billion tons we can still put out before the temperature gets up the two degree increase.

"If we put more than those 750 billion tons into the sky, it will probably get even hotter – by more than two degrees. If we put less, we may be spared some of the woes. We've already put too much into the sky to avoid a one degree of warming. Well, that's roughly how the science is today."

"Well that's simple," said Joe. "So if the real problem is the amount of greenhouse gases we put in the sky, is that included in the international negotiations in the targets?"

"Er … no. Why that is, you'd better ask the international negotiators, Joe. I don't know. They surely have their reasons."

Kelly had been sitting quietly watching them. Then she picked up a Clarets promotion year ornamental paperweight and plonked it into the vase. Water splashed out onto the table. Some petals were knocked off the rose and drifted down onto the table.

"Looks like it suddenly got a bit hotter in your greenhouse," laughed Joe.

"Very interesting," said Igor. "Indeed. My example of the glass vase doesn't cover one nasty aspect of the situation: tipping points. There are a lot of greenhouse gases trapped in the oceans and under the tundra of Siberia and Canada. Trillions of tons. As temperatures increase, the tundra melts and starts to release those gases. It would be as if the water reaches a certain level in the jar, and then suddenly Kelly walks up and dumps a boulder onto it, smashing the lot. We don't want that to happen.

"Today the oceans trap vast amounts of carbon dioxide and suck it down into deep water out of harm's way. But as it gets warmer, that mechanism's already starting to break down. It would be Frank hurling a brick into the vase."

"And as the world's forests die off," he continued, "unable to stand the drought and increased temperatures – they'll stop sucking in carbon dioxide and give up into the atmosphere the carbon they've stored in their trunks, roots, and soils."

"Sounds like a bloody breeze block," said Joe.

"Indeed. It's catastrophic for the vase. If we want warming of no more than two degrees, then humans may put no more than 450 billion more tons of carbon dioxide – or the equivalent of other greenhouse gases – into the sky. If we want to put more than that in, we have to take it out again very quickly.

"Then all those percentages in 2020 or 2050 don't work?"

"They confuse the most important thing – how many tons of the stuff we've emitted in total. So you wonder what they're talking about in their international negotiations."

"Wouldn't it be better," asked Joe, "to be a bit tougher about it? We should just say the world can emit no more than 450 billion more tons of greenhouse gases. Full stop. Then we have a fight about who can emit how much."

He thought for a moment.

"That would help, wouldn't it? The fight, I mean. It might reduce the world's population, too," he said.

31. Efficiency and redundancy

Policies try to be as economically "efficient" as possible.
This might be a mistake.

We played against Birmingham on the Saturday. Frank and Joe and me had been out on the razzle the night before and everyone was grumpy. And Joe was sure he'd lost £20 in the taxi home. We were even more grumpy after twenty minutes when we were one-nil down and could scarcely get the ball into the Birmingham half.

"Bugger!" said Uncle Frank.

"What's going wrong?" cried Joe. "It's a disaster."

Bowyer of Birmingham had made acres of space for himself just inside the Burnley half. A thirty yard lob caught Jensen off his line and bounced off the bar before being hacked away by Jordan.

"Close shave. We're in tatters."

Duff humped the ball upfield to a lonely Fletcher but Birmingham were soon back in Burnley's half pressing for a second. Jordan tripped up Phillips on the right. The free-kick curled into the six yard box, everyone missed it, and McFadden stuck out a leg. The ball rolled towards the line and a desperate lunge by Jensen managed to keep it out. We were hanging on by the finger tips. We held on like this until half-time.

"He's got the tactics all wrong," explained Joe to Igor. "You can see that their midfield's controlling the game … Ferguson's running the show. To win this we need to play, not hoof. That means we need to take Alexander off and bring Cork on, and put Elliott out on the wing."

"Right," agreed Frank. "We never should've played Alexander today. Look at the Birmingham defence."

After half-time the gaffer made exactly the changes Joe and Frank had spoken about. Cork came on for Alexander and Eagles for Blake. Elliott switched to right wing. It made a difference. Burnley's midfield gradually got a foothold in the game and an inspired run down the right by Elliott twisted the Birmingham defence inside out, and Fletcher nodded the cross in.

We almost sneaked a winner in the eighty-third minute after pounding the Birmingham goal for much of the second half. A through-ball by Cork, Eagles outwitted their offside trap, and he lashed the ball over Hart but it ricocheted off the bar into the stands. Still, we'd survived.

"Lucky we had a plan B," said Joe.

"And lucky we had five to choose from on the bench," said Darren.

"Plan B …" said Igor. "If I may have your attention, this is a very important consideration for cutting greenhouse gas emissions."

Darren and Frank groaned. But Doris had some Kitkats and handed them round.

"When politicians talk about cutting greenhouse gases, they talk about efficiency. They want to do this at the lowest cost possible. This is very stupid."

"Stupid?" said Frank. "Surely that's the only thing going for that bloody carbon trading … that it gets us where we want to be at least cost."

"If the Gaffer had made efficiency a priority, would he have brought seven substitutes to choose from? Not at all, he'd have brought three, since that's the most he could use anyway. We need to be very careful with talking about efficiency … It's very different from minimising costs. And it makes us forget to have a Plan B." We couldn't really follow, so Igor tried again.

"Imagine, Frank, you're walking down the high-street. You've a brand-new cashmere jumper-"

Doris laughed. "Frank wearing a cashmere jumper? He doesn't know what cashmere is!"

"Watch it, Doris. It's posh for wool."

"Exactly, posh for wool," Igor chuckled. "Suddenly it starts raining cats and dogs and you need an umbrella.

"Imagine there are three stalls along the street which sell umbrellas. Do you go to all three to check prices and, after finding out which gives best value for money, return to the chosen stall and make your purchase? Or do you go to one and spend five minutes in the rain bargaining with the umbrella seller, saying that you can always take your custom to the next shop along the road? Or do you just buy the first umbrella you find and preserve your new jumper? Remember, it's raining very heavily."

We all thought about it.

"Well I'd just buy the first umbrella I see," said Uncle Frank.

"Me too," said Joe. I agreed with Joe and Frank.

"Exactly. Even though it's not the cheapest umbrella, you buy the first one you see because in the circumstances the heavy rain is very likely to be catastrophic – for your jumper.

"Well, with greenhouse gas emissions, we might be already in the thunder storm. We don't know how much time we have. We should think twice about being too clever about spending a lot of time to work out the lowest cost approach. We should just do something which works."

"That's what I've been tell you all along," said Frank. "We've all these clever bastards in universities and political parties and research centres ... do nothing but fly about the world to conferences and congresses ... taxpayers money ... carbon this, carbon that ..."

"You've set him off, Professor," warned Doris.

"Shut up, Frank," said Joe. "We get it."

"We don't have time to find the perfect and cheapest solution. We must get away from the notion of efficiency. We simply don't have time for it. We must do things which work.

"Then," continued Igor, rushing a bit to keep our attention, "there's the question of redundancy, in many ways the opposite of efficiency. We're familiar with redundancy in the sense of something which bankers are given. But I'm thinking of another kind of redundancy."

He continued. "For today's match the Gaffer probably had two or three tactical scenarios. He didn't know which he'd need to use, but still he had to be ready with them. He put seven on the bench. Again, he knew he'd use no more than three, but had seven in case.

"Redundancy is about deliberately building in back-up systems in case a main system doesn't work – extra tactics, extra players. We can't predict the future. We don't know what'll work and what won't. So we have to have plenty of redundancy to reduce the risk of things not working out at all."

Joe nodded. "You need your Plan B. And possibly Plan C. Or even D just to be sure. Imagine they'd not brought Jack Cork today. We'd have lost, simple as that."

"And in climate policy we have don't even have a plan B. We hardly have a plan at all," said the Professor.

32. Resolution

To make cuts of 80% or 90% we have to get to every nook and cranny.

I needed a break from it all and a good helping of steak and kidney pie. The insurance money from the car crash in the summer had finally come so I took my new Ford Focus for a drive up to the Lakes. I was back on Wednesday evening just in time to check that the Professor hadn't burnt the house down and then dash over to the Bolton game.

We succumbed in driving snow. It was all OK until half-time. Then things started going wrong and we let in two goals in quick succession which finished off the game.

Joe was ready with an explanation. "You see, I've been thinking about this. We were focussing on the big game on Saturday against Chelsea. It's tactics, you see. You've to focus on the important things. Bolton, I mean, it's only a local side. It's – "

Frank slapped Joe on the back of his head causing him to spill his pint.

"What'you do that for?"

"You're talking crap. We should focus on the ones we can win, not the one's we haven't a chance at, you daft twat."

Igor laughed. "Your excuse, Joe, is nonsense. Every game

counts, every point counts. You get the same with climate policy. People say: oh this and that is insignificant. Like: having baths. They say we shouldn't be focusing on that. But the trouble is, if you look at the numbers there's practically no source of emissions which is insignificant."

Joe had caught sight of Beanie and Chopper waving to them from the other end of the bar and moved to follow Frank along the escape route. But Igor raised his glass and Joe missed his chance. Igor smiled to him and gestured to him to sit down again. Then he explained that emissions come from out of every single nook and cranny of society. There's no single place you can point to and say: "If we can tackle that, then we've broken the back of the problem." The problem doesn't have a back.

He seemed to go off on a tangent and said that when you peel carrots, a few smooth slices of the peeler will have the carrot done in seconds. There's a process and a rhythm and you can get through a bag of carrots quickly. That is not like the problem of climate policy, there's no systemic solution. Climate policy is like peeling a Jerusalem artichoke.

Joe didn't know what a Jerusalem artichoke was, so the Professor drew one on the back of his beer mat. "You see, there are no short-cuts, no quick slices which'll cover 60% of the surface of the vegetable. You have to get your little knife into all

the irregular knobbly shapes if you want to remove the peel and retain a reasonable amount of the artichoke."

"Look at the Germans, for example," said the Professor stepping with agility from artichoke to sauerkraut.

"I'd rather not," replied Joe. "They're cheating buggers, and their birds are —"

The Professor meant the report which the Germans sent to the UNFCCC in April 2009 listing all its emissions. "Ah, right," said Joe. The Professor handed him a list of the top ten sources of Germany's one-billion-plus tons of emissions.

"Now, say we want to reduce emissions by 90% to be safe. We start with heat and power production and then move steadily on to transport and residential energy use.

"Assume that we can make all heat and power generation carbon-free. All power will be from sun and wind and rain and nuclear. All the gas and coal fired power stations will be shut down or equipped with spanking new carbon capture and storage technology. It was hard work, but we managed it. So we got rid of 340 million tons and reduced emissions by 32%. Still more than 660 million tons per year to go. What's next?"

Joe looked down at the list he was holding. "Er ... road transport. It's 13% of emissions. Well, you'd think that Porsche and Volkswagen and Mercedes and Audi could sort something out," he remarked.

"It's either that or Fritz can hop on his bike. Imagine they do manage to eradicate all cars and lorries that run on petrol. That is all commuters, all truckers, all cabbies, ambulances, vans-"

"What, even the fuzz?"

"The fuzz?" asked Igor.

"Coppers, policemen."

"Yes, they all go on to electric or pedal power. This vast transformation for just another 13% of reductions.

Igor said that to cut emissions by 90%, we would need to get rid of all the next thirteen categories of emissions. Then he said we should imagine there's some slippage in the great plan to make the power sector free of emissions. Say we only get 80% of the way, and end up 70 million tons short in the power sector. To get the extra 70 million tons of reductions we'd need to eliminate emissions from lots of industrial and agricultural activities: the production of carbon black, of ethylene and ethylene dichloride, of styrene and methanol, of catalysts for refineries and of fertilisers. Emissions of methane from waste sites would have to stop, and the same for cows and other farm animals – back and front. Farmland would have to stop emitting carbon dioxide and nitrous oxide. Leaks in gas pipelines would have to be stopped up. The iron and steel industry would magically go to zero carbon, as would lime production. All that extra effort needed if you can't make electricity production carbon free.

"Then," the Professor continued, "if we can't get Porsche and supertrucker Willi Betz – the Eddie Stobbart of the Autobahn – to play ball, and say we only cut vehicle emissions by 80% instead of 100%, then to compensate for that 29 million ton shortfall we need to eliminate emissions from the manufacture of adipic acid, ammonia, from coal mines, from peat extraction, from aluminium and magnesium production, from gas compressor stations, from domestic refrigeration, commercial refrigeration, mobile refrigeration, air conditioning systems, foam production, fire extinguishers, aerosols, solvents, switching gear, insulated windows, sports shoes, AWACS maintenance, welding, and the manufacture of optical glass fibre and photovoltaic cells. The armed forces have to go carbon neutral and that includes the Luftwaffe.

Joe laughed. "Nice one, bloody Red Baron in an air balloon."

The Professor carried on. "So, any slippage in the main categories of emissions means we need policies to cut emissions across all aspects of society. Let's say we can only cut emissions from home energy use by 90%. Well, after that we bump into a brick wall. There's practically nothing left to reduce except for the candles in Cologne Cathedral. We may need to keep them burning."

Igor let Joe go and join the boys. We drank our pints quietly. He turned to me and said this isn't about whether you put bouncy solvents in Adidas shoes. Even if all energy production and transport go zero carbon, there's scarcely room for emissions from the current technologies of iron and steel, cement, concrete, glass, bricks and ceramics. Especially if we need some wiggle room. We'll have to do without them. Hold on to that Meissen collection.

Then he said there's scarcely room for the production of ammonia. It's the raw material for making nitrogen fertiliser. If you've ever seen a German eating, you'd know how important nitrogen fertiliser is.

"And now," said Igor, "that we are on to the problem of agriculture. In the German case – but it's much the same elsewhere – about 16% of emissions relate to land use. Every time we disturb the soil it expresses its discomfort with a guff of greenhouse gases. That 16% is practically untouchable. Organic farming can help in some cases, but we've little room for error. You can't cut by 90% and leave 16% – the numbers don't work.

"Unless," he said, "the Germans really do stop eating so much sausage and salami."

"Or just stop eating so much, full stop," I suggested.

"Or just stop eating, full stop," he replied. We finished off our lentil bake, the Professor with more relish than me.

"So you see, all this shows that any slippage in cutting one of the large sources of emissions means that dozens of other industries and activities have to make additional reductions. A shortfall in one industry cascades across many others, like a fall of snow in a mountain triggering avalanches across the valley.

"We won't get 100% reductions in power production and transport and domestic heating. It means that every other category of emissions will need to be drastically cut as well – hundreds or thousands of industrial and social practices need to be changed, sometimes fundamentally. It will touch every area of our lives, however small."

33. The wrong comparison

Policies which count the cost of losing our current infrastructure are missing the point – the whole purpose of the policies is to replace that infrastructure.

Things got worse. That Saturday we were hosts to Chelsea in an early game and crumbled. We let in four goals in the first half and then Chelsea did passing practice for 45 minutes.

Another thing crumbled. It was Joe's new year resolution not to moan if we lost. "We just weren't in their league. It's hopeless. We're never going to stay up if we just collapse like that. We were nowhere…" It went on for some time. Once he'd calmed down and was sat in the Bridge with a tankard of Bank Top down his throat, the Professor tried to reason with him.

"Joe, look here. You're unhappy about it because you're comparing with the wrong thing. Of course a game against Chelsea shows up your weaknesses. They're a fine team. They're fourth or fifth in the league, aren't they?" Far too high up for my liking, but while we're on that sensitive topic of Chelsea, I'd heard on the grapevine that Mr 50k – the bloke who'd nicked my bird – wasn't all that he seemed to be so I lived in hope.

Igor was still talking. "It's most important to choose the right thing to compare yourself to. Most important. I can give you an example from my very own field, if you like."

Before Joe and Frank could agree, he fixed them with a stare and began.

"When people are looking at what to do about cutting emissions, they look at how much it costs to cut them. In order to get a feeling for whether it's expensive or not they compare them to the current economy. They think 'We've got a big economy with lots of factories and power plants and lorries and planes and people keeping busy and we want to cause as little disturbance as possible to it.' So any policy measures must be as low cost as possible in order not to cause any disturbance.

"So far so good," said Joe.

"Unfortunately they're comparing the costs with the wrong thing. The things we've got to do are such big changes that the economy we're comparing with won't be there anymore. So it's irrelevant. See what I mean?"

"Mmm…. not really."

"Imagine your world is an egg. You want to make an omelette. Do you compare the cost of breaking the shell with the value of the egg? Or with the value of the omelette?"

"Bloody hell, Igor," laughed Frank. "You're so deep, I need a bloody bathyscope to keep up."

"Or down," suggested Joe.

"You see," said Igor, "We aren't talking about some little scheme on the edge of the economy. This isn't fine-tuning. We're talking about replacing big chunks of the economy – well over half our economic activity isn't sustainable. Once these chunks have gone, we won't have the same economy. Our industries and their products consume stupendous quantities of fossil fuels. All will have to be fundamentally changed in a low-carbon economy. From the point of view of society, the technology of today is junk and has no value. So why would we worry about the costs of dismantling a machine which we know can't survive?

"It's not logical. If we've agreed we've to get rid of it, surely we should write it down to scrap value." Igor paused.

Then he said, "The so-called costs of climate change policies are the costs of switching to a sustainable economy within which mankind and other species can survive. These aren't costs, they are investments. Therefore they should be compared with the benefit of obtaining a new economy which works.

"Things which seem to cost us money now in terms of today's economy, are actually helping us move more quickly to a sustainable economy where we'll create wealth, wellbeing, happiness, satisfaction, or contentment by other technologies and behaviour. We should spend lavishly and boldly on this goal."

"I'd spend lavishly and boldly on any goal just at the moment," said Joe. "As long as we score it."

34. The wrong numbers

The UN's official records of emissions look at where the emissions come from, not why they are there. An inventory of "why's" might be more useful.

Back at home a bit later, Joe was reviewing match stats with the Professor. "Look at this, since mid-December – lost, lost, lost, won – that was the Newcastle game when they missed the penalty – drawn, lost, lost, lost. Five goals for, seventeen against. Four losses to nil. No goals scored in the last three games."

"It's a bad run," said Igor with sympathy.

"The numbers tell it all, don't they?" said Joe. "In black and white."

Igor looked at him, surprised. "Do you think so?" he asked. "I'm not so sure they tell us much at all. They don't tell us what went wrong, do they? They don't tell us who played well and who played badly? They don't tell us why. They don't tell us about the team spirit in the dressing room. They don't give us the gaffer's pep talk at half time. They don't tell us about where the weaknesses are. They tell us what, but not why. Less than half the story."

Joe laughed. "You've always got an angle haven't you?"

"Not at all. It just happens I was thinking about the very same problem. Remember that long German report from April 2009? 565 pages long. Well the UK one from April 2008 is 568 pages long. And the French one? What do you think?"

"No idea. How could I know that, Prof?"

"It is 1,196 pages long, longer than the German and UK efforts put together!" exclaimed Igor.

"These are outstanding works of art and science. They elaborate in exquisite detail the median nitrous oxide emission factors of dust combustion lignite plants larger than five megawatts and smaller than fifty megawatts in the new German states and they tell us the emissions of methane from calves, heifers, fattening bulls, suckling cows, and stud bulls between 1990 and 2007. The German report alone lists forty-five authors."

"Jesus," said Joe.

"Bloody waste of time," said Uncle Frank.

Doris came in with tea and a plate of scones.

Igor wiped the jam off his face and continued. "So these reports provide us with encyclopaedic understanding of the sources of greenhouse gas emissions. Yet in a sense they tell us nothing. Rarely has the work of so many diligent, earnest and intelligent people been so futile."

"Meaning?"

"He means these books are a waste of time," explained Frank.

"Exactly," continued the Professor. "If we want to do something about greenhouse gas emissions rather than just admire their anatomy, we need to know not the sources of emissions but the reasons for the emissions.

"The United Nations should prescribe a national inventory based on 'whys' and not 'whats'."

"You what?" asked Joe. "Why?"

"The national inventory should be based on the causes of emissions, not the mere sources. They should ask each country why we have these emissions. Remember what I said: we need to know the motives behind emissions, the psychology behind them, not what machine emits them.

"This inventory would be more difficult to compile. It may be satisfyingly impossible. It would require judgment, insight, and some wisdom. It would need the courage to speak out of turn. It would demand greater skills and deeper humanity than the current inventory which is an exhaustive exercise in taxonomy and counting molecules.

"Yet the inventory of whys would be far more useful. It'd tell us how many emissions relate to poorly insulated homes; how many emissions are caused by people unwilling or unable to walk for twenty minutes; how many emissions happen under the influence of advertising; how many emissions are attempts to console unhappiness; how many emissions are escaping regrets of the past or fear for the future; how many emissions are the soulless grind of commuting; how many emissions are payments in brown envelopes; how many emissions seep out as a result of eating meat."

"So they're barking up the wrong tree?" said Doris.

"Or just plain barking," said Frank.

"Completely. They're only interested in recording numbers and not understanding," said Igor.

"Like nerds who know all the stats but never go to the games," said Joe.

"So collecting all those numbers ... it won't help them get to the heart of the problem, will it?" said Doris.

"Course it won't" said Uncle Frank. "Like I said, it's a waste of bloody time."

35. Winging it

Igor moves on to questions of replacing our energy intensive infrastructure.

Frank mellowed after his cup of tea and fell into a more philosophical mood. "The fact is we've to rebuild the entire team if we're serious about staying in the Premiership. The squad's too small, there's too many injuries, we need a striker who can put in twenty a season, we need an anchor in defence, and we need a midfielder who wins matches ... and we need a quick winger ... like Walcott or Valencia: we need some big stars ... Personally, I'd start from scratch."

"We'd need millions for that, Frank," said Joe. "It'd take an Arab or a Chinese billionaire. Not many of them interested in Burnley. Running a takeaway, you're just not in the right league. Plus ... there's not so much time, is there? There's three days til the transfer window closes. With all the money in the world you couldn't build a new team now. The only deal we're in on is Tevez. And even that is touch and go."

"Carlos Tevez. Our great white hope," said Frank.

"It sounds rather like a pipe-dream to me," said Igor.

"Well that's where you're wrong, actually," said Frank. "Chopper's mate says there's this Russian consortium and they're ready to back us. It's definite. The board's flying down to London tomorrow morning to finalise it."

"You sound just like a politician," Igor replied. "Full of dreams and no practicality. It will never happen."

"Oh yes, Igor. How do you know that, then?"

"I know that because it sounds just like the fantasy-world stuff which goes on in my field. They're saying demand for energy will carry on going up and at the same time we have to stop emitting carbon dioxide. The only way both can happen is if we change all our factories and all our houses and buildings and all our cars with new equipment which does the same thing but doesn't emit greenhouse gases."

"Right, so we've got to rebuild everything," said Frank. "Nothing dreamy about that. It's solid business. How long have we got?"

"Twenty, thirty years."

"Thirty years? That's bloody ages. We've got three days to close the Tevez deal. Transfer deadline closes on Tuesday night."

"It's difficult to know if thirty years is plenty of time or not … they need to replace 12,000 fossil-fuel-fired power stations; plus the three new power stations which are completed each week – that'll be another 3,000 by 2030. Then there are at least 2,000 cement plants, 1,000 steel mills, 650 petroleum refineries, and perhaps 10,000 chemical plants. All emitting something like 20 billion tons of carbon dioxide each year.

So we have to replace or retrofit something like 30,000 industrial facilities in thirty years. Then there are 600 million cars running on petrol in the world. And in England alone there are 22 million houses which need rebuilding."

Frank nodded, weighing up the numbers. "Right, all in thirty years, plus or minus."

"But I thought you said we've to stop emissions growing any more within the next five years," said Joe. "Doesn't that mean they've got to do it now, not in thirty years time? Won't that be too late?"

"They'll have to do a good chunk in the next five years, definitely," said the Professor.

"How long's it take to build a power station? Six months or so?"

Frank laughed. "Not quite," said Igor. "The trouble is that in heavy industry the equipment is replaced very rarely – once every twenty or thirty years or more, and technology changes very slowly. The basic technologies of power generation, or making cement and steel – they've barely changed in the last century.

"These aren't industries like computers where new things come out every week and a computer's replaced every two years. These industries are technologically dormant and move like sloths.

"The politicians are expecting them suddenly to spring to life and dance the cha-cha-cha like media and information technology!"

Joe was thinking. "Well I guess if that's what they're planning, they must be really confident they can do it," he said.

Then Frank started up, as if something had twigged. "It's bollocks. It's just bollocks," he said. "They're just winging it."

"Winging it?" we asked.

"Yeah. And I'll show you why, too," replied Frank, tapping into his Blackberry.

36. A new stadium

Why it is not so easy to rebuild our industrial infrastructure by 2050.

We piled into Frank's BMW and he sped off through the Saturday afternoon traffic to Briercliffe. He reached an industrial site to the north of the town and turned in through some gates onto the forecourt of a ramshackle warehouse. He'd been on the phone to his mate Chopper on the way and the Volvo SUV arrived just as we did.

Frank waved his hand across a wasteland of office blocks with shattered windows, empty warehouses and tumble-down factories looming in the dusk. "Turf Moor is dead. Long live Turf City. Me and Chopper. This is where we're going to build Burnley's new stadium.

"It'll cost £50 million quid. It's going to take three years at least. And that's why they're winging it."

Igor and Joe didn't follow.

"A little bit of real world, Professor. It's a big project. A very big one. And that's the issue. It's a very long process. Long and complicated. And that's just a football stadium."

I asked Frank to explain so he started listing all the things needed to get a project to happen. You get an architect to put forward a concept design. You need to find the site. You need to be sure that you can buy the site. You don't want any hazards on it, no hidden pollution."

"Then you have to get rid of any rare birds or plants, so they don't find them," added Chopper.

"And then you have to get the local community on board. Should be OK with a football stadium. Might be harder with a power station – which sort did you say we need?" asked Frank.

"Wind, nuclear or coal – in some combination," replied Igor.

"Right," said Frank. "So every community in Britain's pretty much against a wind farm, these days. Spoils the view, doesn't it? Knocks a few grand off your des res in Cumbria. So that might not be so easy. Then with your nuclear, yep … I reckon pretty much every community in Britain'll be against that one, too. Now with your coal…"

"Yeah, with your coal," said Chopper. "Probably …" he paused thoughtfully. "Probably every community'll be against that one, too. Anything else?"

The Professor gestured empty hands.

"Look," said Chopper. "If you want to get the local community on board, you have to do a deal with them. Take out the noisy ones with a little sweetener. Sugar the pill. Silence has a price.

Know what I mean? Then you throw in some crappy housing for the homeless or knock up a shed for the local comp."

"It takes time. They can get argumentative. Fucking nimbies. You have to soften them up. Nowt nasty, mind."

"Just like Chopper says," confirmed Frank.

"Then once you've got the locals tied up, you'll have to cut a deal with the council." Chopper walked away, lit up a cigarette with his Zippo, shielding it from the wind. He turned back to them, heavy set, black leather jacket, silhouetted against the weak winter sun on the horizon. "You know what I mean, right?"

Joe didn't.

"Look, there's something I want and there's a little man on a committee who can give me that something easily. But he might not want to give it me. So I have to help him want to give it to me, don't I. Scratch, scratch."

"But that wouldn't happen with power stations and stuff, would it?" asked Joe.

"Probably not," Igor said. "Which is why so many wind farms don't get approved."

"Well if you don't bloody do deals with the planners, of course you're not going to get stuff through. Greenies don't like getting their hands mucky," said Chopper. He flicked his cigarette end into the grass.

"Then once you have your indicative planning permission ..." began Frank.

"Hold on," said Joe. "How long's that all take?"

"What? From the first idea?"

"Yeah."

"Phew..." Chopper lit up again. "Concepts ... local residents ... deal with council ... a year or more perhaps," he said. "Then you've to do real work. Architectural and technical plans, deal with your suppliers, putting together the finance."

"Then you're starting to spend real money," added Frank. "Architects, designers, engineers and consultants, the whole finance side, your financial advisors, your taxman, your project managers.

It's a good two years work to put all that together. Easy."

"And all that time you're working the community, working the council," said Chopper.

Joe asked the Professor how long this would take for a power station.

"Three to five years for a conventional power plant. Five or more for something new like a coal plant with carbon capture and storage ... ten years for a nuclear plant. All the studies you have to do – huge public consultation, traffic surveys, environmental impact assessments, negotiations with the energy authorities, deals with suppliers of coal or uranium, complex deals with lenders, banks, investors, syndicators ... involving thousands and thousands of pages of documentation ..."

"No health and safety?" asked Joe.

"And that," said Frank.

"You'd need a public enquiry for something like that," said Chopper. "For a little housing development, a few bob here and there can sort it. But these days for a big'un, you're going to have a public enquiry."

"And how long can that go on for?" asked Joe.

"Years," said Chopper.

"And then?"

"Well, once there's a decision, the lot that lose go for a judicial review."

"Judicial review?" said Joe.

"Yeah, they bring law lords in."

"Bloody hell. It must go on for years," said Joe.

"Eventually one side gives up and writes off their losses."

"And they have to do that each time there's a new power plant?"

"Any big industrial facility. Power plant, cement plant, steel plant ... Course sometimes local council approves, sometimes the government."

"Like all the stuff which the Professor says has to be replaced?"

"Sure thing," said Chopper.

"So what happens next?"

Chopper explained that once the whole thing has been finally approved by the authorities, and when you've got all the money lined up, then you can start building the thing.

"How long's that take?"

"This stadium and shopping centre complex? It'll take three years to build."

"What about a power plant?" asked Joe.

"Power station," laughed Chopper. "You're talking five years for coal. Ten for nuclear."

"But once you've got your permits, then it goes pretty smoothly, right?"

Chopper laughed again. "Yeah, until things start going wrong. Nothing goes to plan, mate, does it? Disputes with suppliers, problems with getting materials and parts, subsidence, a strike, an accident here and there, burst pipes, logistical mix ups, another go-slow, miscommunications, subcontractor sends stuff to the wrong place, shortage of engineers, pump breaks down, worker goes missing, finance has to be reworked, tax laws change, … thousands of things go wrong."

"And that's even when the technology's well known?" asked Igor.

"Sure, I mean power stations are a bit more complicated than a housing estate. Of course there are delays and stuff. It takes ages to get them built."

"And this is just Britain. What about getting things done in developing countries?" asked Igor.

"They'll have their work cut out." Chopper stubbed out his second cigarette and pointed the key at his car. "Gotta take a man out to tea. Know what I mean? Tara lads. Oh, and good luck with your penguins, Professor."

Chopper sped away, headlights cutting across the piles of rubble and nettles.

Joe was perplexed. "If it takes so long to build all this stuff, why don't they start now? If they started now they perhaps they could be done by 2030. Or 2040." But none of us could answer that and it was too cold to be hanging around in Briercliffe.

37.   The Great White Hope

The problem of carbon capture and storage.

On Monday evening Doris was out and we had to make do with a pub supper. We were finishing off our steak and Guinness pie when Frank got a call from Spike that the Tevez deal was definitely on. A mate of his had told him that he'd been talking to someone who was close to the fire. The Russians had backed off but there was a new consortium of investors – Kazahk metal traders – and they were ready to bankroll it. The Chairman had been back in London all day for talks. The lawyers were ready with the paper work. It was all last minute, but it was definitely going to happen. Spike was going to ring back if there was anything new.

"Something's bugging me," said Joe, once Frank had relayed the news. "Not about Tevez. About building all them power stations and stuff. If it's so urgent, why aren't they getting on with it?"

"A hundred reasons," said Igor. "Too much uncertainty. We're bang in the middle of an economic crisis. It's hard to make investments these days. And as long as the political cycle lasts just four years, you can never be sure that the laws and regulations will outlive the government."

"And without building all this stuff, there's no way the human race can survive?" asked Joe.

"Not in the premier league," said Igor.

"So what are they going to do?"

"There's a brand-new technology; it's called carbon capture and storage. CCS. It's our Great White Hope," explained Igor.

"Relying on new technology? To save the planet?" Frank snorted. He said that was bollocks for a start. Basic risk management in any building project – don't try out fancy stuff.

I asked how it works.

"They capture carbon dioxide emissions at the coal-fired power plant, or an iron and steel works, or some industrial facility … and then squirt it into a hole in the ground where it will stay for thousands of years."

Darren laughed. "That sounds just like Tevez. He's a sort of technology for getting the ball on the edge of the six yard box and squirting it into the net. That is so what we need!"

"I wish I could share your enthusiasm. The difference between Carlos Tevez and Carbon Capture and Storage is that Carlos Tevez has been scoring goals for years at the highest level. Carbon Capture and Storage technology doesn't even work yet. It's like a lad who is still at the academy school. Aged nine and a half."

"Jesus. So when's it going to be ready?" asked Joe.

"The experts expect that it'll be commercially viable by 2025 or 2030."

"By which time Tevez'll be retired."

"Exactly," continued the Professor. "And by the time carbon capture and storage technology works and is affordable, we'll have ten to twenty years to implement it globally. If you need to put this technology in somewhere between 5,000 and 10,000 plants, you'd need to do two a day worldwide without stopping on Sundays. And remember the difficulties Frank told us about building large industrial facilities. Two a day," repeated the Professor.

"You're joking!" said Joe.

"It's not just every coal-fired power plant in the world – if you can retrofit it, that is. Cement plants and steel mills and chemical plants will also need Tevez technology."

"There's no other way of making that stuff without carbon dioxide emissions? There must be something they can do?"

"No. They're just waiting for carbon capture and storage to solve their problems."

The barman came and removed our plates and took an order for four sticky toffee puddings and a small cheese selection.

"I told you it's a waste of bloody time," grumbled Frank. "If these industries are just counting on some new technology, they're stuffed. By the sound of it, it's not going to happen."

"Ok," said Joe. "So what's needed to get this all up and running? Say it's 2030 and the technology works."

"A long shopping list," said Igor. "You need to decide which authorities are responsible for regulating this complicated new technology. You need insurers."

"For a brand new technology no-one really understands. Yeah, I can see a lot of insurers volunteering for that," said Frank.

The Professor continued. "Then you need salesmen who know what they're talking about. Commercial references are needed. It's a Catch-22 situation for new technology: without references it's hard to sell; without sales it's hard to get references.

"Then you need engineers who can design and build the stuff. Where will these engineers all appear from so suddenly? How will the professors at engineering schools learn the practicalities quickly enough to be able to pass the knowledge on?"

"Right," said Joe, "when the paint isn't even dry on the technology yet."

Frank snorted. "Engineers? I can't get a decent engineer to design a drain system, so you can forget about your fancy Carlos Tevez technology. This country doesn't produce engineers any more. Maths and physics is too hard for the kids these days."

Igor persevered. He said that you need suppliers who can deliver the equipment on time. This time we won't be able to afford delays. "Remember, one plant a day. Someone has to sort out the supply chain pretty effectively."

The pudding arrived.

"Then the permitting authorities have to get up to speed and – "

"Work out how much in kickbacks to ask for," interrupted Frank.

"But like, not here, though, right?" said Joe.

Frank smirked. "No, no, no. Not in Britain. Course not. But in most of the other hundred countries where this stuff has to be put in."

"Then – " asked Joe.

"Then you need to persuade investors to come on board."

"Investors," laughed Frank. "Investors these days, they'll take as much risk as the stuntman in Tellytubbies. I don't think they'll be rushing to fund this stuff."

Joe asked where they were going to bury it all.

"Under the North Sea," replied the Professor.

"Right, so they're going to build massive pipelines from every power station across the country and into the North Sea, are they?" asked Frank. "Right, using highly experienced oil engineers from the Gulf of Mexico. Goodbye reality."

"But haven't they done if before," said Joe. "With north sea gas? Like when we were lads."

"Perhaps so," said Frank. "But look, if they won't let you put up windmills, they're not going let you cover the country in more pipelines, are they?"

"You know something," said Frank after a while. "You've forgotten the most important thing."

"Really?"

"Well, the numbers have to add up, don't they? I mean, the price of electricity has to be high enough for it to be worthwhile for someone building all this stuff. But if the technology is new no-one will have a clue what the right price should be. It will be years before the regulators and the operators agree on prices which are at the right level.

"Either it's too low and no-one makes any money from it. Or it's too high and they're laughing all the way to the bank. Either way, it won't work."

Joe gave out a moan. "And all this has to happen in over 100 countries for us to stand a chance?"

"That's the plan," said the Professor. "That's the Great White Hope."

38. Write-offs

Policies to cut emissions should result in huge write-offs of unsustainable power stations and factories, leaving the banks with a nice problem.

The Kazahk money never materialised and the transfer window closed on Tuesday midnight. The Tevez saga fizzled out. The next game was against his old club, West Ham.

We stood on the terraces in heavy coats. The wind scudding off the Pennines blew in a reminder of what winter should be like.

Burnley played a great game with some showy football. Joe was chuffed, and believed that surely now the Professor would sense the spirit of the game.

"Look at that! How they beat the offside trap. Now that's beautiful football. See how he timed his run perfectly, just as the ball was released. Got in behind the defence. They were all wrong-footed. Beautiful."

But Igor didn't get it. "I can see the timing but not the beauty. It all happened in such a flash, and now it's gone. All that remains is a statistic."

Joe sighed. Igor had developed a powerful intellectual understanding of the game. But it was all in black and white and numbers. There was no twitch of passion for him. Try as he might, Joe couldn't get him hooked.

Joe couldn't understand how a through ball like that by Cork and a perfect finish by Paterson could leave Igor cold. Still, it was a precious point. We now had 20. We were just starting to dare to hope.

"With so much hope, you could be a politician," remarked Igor as they settled in the nook in the Bridge. "They place all their hope in technology. Not just white hopes, dark ones as well – hopes of uranium, concrete and steel, hopes of great reflective structures in the sky, or engineering the oceans, of filling the stratosphere with acid ..." He shuddered.

"But nature has a cruel sense of humour. By cruel coincidence all the things which we need to replace – power stations and things – have very long lives. Power stations last thirty to sixty years. So do aluminium smelters, chemical plants, paper mills. Glass, cement, and steel factories are built to last up to a hundred years.

"Built to last, eh?" joked Frank.

"Ironically, 'built to last' is no longer a good idea. This isn't the throw-away world of mobile phones or laptops. IT technology spreads quickly because you write off your mobile phone after a year and get a new one. It's a fashion statement and you want to replace it quickly. Cement plants aren't fashion statements.

"Because it's so long-lived, this equipment is scrapped and replaced only rarely. Most of the power plants standing today will still be operating in 2050 assuming they aren't forced to close down sooner."

"But if they're not forced to close down, doesn't that mean that they'll carry on polluting?"

"Stands to reason, Joe," said Frank. "If they carry on operating, then they carry on polluting."

"But … they're operating today and making money, right?" asked Joe.

"Yes, that's right."

"And you're saying that they'd naturally carry on operating for another thirty or forty years."

"Yes."

"But doesn't that mean that …-"

Frank jumped in. "Bloody hell! What's he's trying to say, Igor, is that if these plants are operating now … and they'll have to be shut down before they've reached the end of their life … then someone's going to take a hit. A massive hit. All those plants are on people's balance sheets, right? And they're going to be forced to close. That means big write-offs. And if they've been funded with debt. So the banks are going to have fun all over again."

Igor was puzzled. "Then how on earth do you get these companies to change technology in time?"

Frank winked. "First rule of business, Professor. People only do things if they have the motivation."

"What about Tevez, then?" asked Joe. "We had motivation to buy him. We still didn't."

"Motivation … and, the right price," said Frank.

39. Price of a ticket

*Igor now moves on to the question of carbon pricing – making
people and companies pay for their emissions.*

Joe was livid. "A thousand bleeding quid for the train?"

"Just over. For the four of us," said Doris.

Joe and Doris were sat at the computer looking for tickets to
get to the first-leg against Paris Saint-Germain later in the month.
The glamour of Paris made the trip a must.

"It's taking the mickey. It's four hundred with Jet2. A hundred
quid a head. And it takes longer by train. Bloody crazy."

"What if we hitchhike, dad?" asked Kelly.

"Oh yeah, we'd look right bloody Charlies four of us on the
M65…" said Joe.

"Waiting for someone who just happens to be driving to Paris,"
added Darren.

"Look, Prof," said Joe, "I'm sorry, but we can't do this one by
train. Time-wise it's fine, but look at the cost."

Igor smiled. "Let it be my gift. I might not yet feel the passion
for football, but my betting is becoming increasingly successful. I
have analysed the prospects for the Middlesbrough game and I
have a clear view of how it should proceed. I daresay I'll win back
the extra £600 over the next few games."

Darren whistled, impressed.

Doris and Joe looked uncomfortable but Igor would have none
of it. "It's my pleasure," he insisted and then: "As Frank would
say, you need the motivation. An extra cost of £150 per ticket
doesn't motivate anyone to use the train."

Darren asked why don't they make it more expensive to fly and
cheaper to take the train.

"They're trying to, but not very effectively." Igor laid out the
problem. There is an idea that to get people to switch from one
thing to the other – like switch from going by plane to going by
train – all you have to do is to make, say, travelling by plane more

expensive. If the plane tickets were at least £150 more expensive, then people would be more likely to go by rail.

Frank grunted that if they did that, we'd be going nowhere.

"But let's say they did charge an extra £150 per person for the flight," said Igor. "The way they want to do this is to charge you for the carbon emissions. So you'd pay £150 more for the flight because it emits, say, a 170 kg more to go by plane."

"Makes sense," said Frank. "If they want it, you have to make the buggers pay for it."

Darren got out his phone and flicked through to the calculator function. "Right, if it emits 170 kg more to fly to Paris than take the train, and you want the flight to cost £150 more to make it the same as the train journey, then you'd have to put a price of … Hold on. If 170 kg cost £150, then one ton would cost £882, right? You'd need a carbon price of £882 per ton."

"Well that's not going to happen, is it?" said Frank. "It would put every power station out of business."

Igor laughed.

Darren asked what the cost of emitting carbon dioxide is today.

"It ranges between £10 and £15 per ton of carbon dioxide in the carbon market," replied Igor.

Darren got out his phone again. "Say £13 a ton … and we said it's 170 kg more to fly than take the train. So if they made you pay for the carbon emissions of your flight today, you'd pay an extra £2.20, right?"

"£2.20!" laughed Joe. "That's not going to stop anyone flying is it? It's a bloody pint."

"You're right," said Igor. "But you'd be surprised how many people think that putting a price on carbon is the solution to all our problems. Politicians, journalists in leading financial newspapers, authors, academics and economists – they glibly say that once we have the right carbon price then things will sort themselves out. People will stop wanting to buy carbon-intensive products because they'll get too expensive. And they'll start to buy low-carbon products which'll be comparatively cheaper.

Gradually, as carbon prices go up, the carbon intensive economy will be replaced by the low carbon economy, just in time for 2050."

"This is another of those nice dreams, isn't it Igor?" said Frank.

The Professor laughed again.

"But it's kind of true, isn't it? If the price of beer goes up, people drink less," said Joe.

"Do they? Do you, Frank?" asked Igor turning to the man with the money.

"Course I don't. Beer's beer. It's a necessity."

"Exactly," agreed Igor. "In theory, if you make it more expensive for industry to emit lots of carbon dioxide, they should start to look for ways to cut their emissions – improve their technology or replace it with cleaner technology or change their products to ones with fewer emissions."

"Like they should tax foreign players to encourage more investment in local lads," said Joe. "If it were more expensive to bring in the big names from Brazil and Spain, then we'd spend more on the juniors. Stands to reason."

Igor agreed. Putting a price on carbon also makes it more expensive for customers to purchase carbon-intensive goods and services. After a while, they start to feel the pain and will (in theory) switch to goods which take less energy to produce or use. Like switching from the plane to the train.

The third thing which pricing does is gives signals to investors. It's a way of policy makers telling investors what conditions to expect over the long term.

Joe shook his head. He didn't see how they know how much to charge for emissions. "What I don't get is how do you know if you should be paying 150 quid for your carbon or two quid twenty?"

"They don't know. In some cases they charge for emissions by having a carbon tax. There the economists work out a number – then they wait and see what impact it has. Another way of charging for emissions is to have a trading scheme as they do in the EU. They set a cap for all heavy industry – a limit of how much carbon dioxide can be emitted in total, and companies in the trading scheme can buy and sell allowances to emit carbon dioxide according to their needs. A market develops for the allowances and gives them a price. Just like other things which people buy and sell – coal, chocolate, beer – they all have a price determined by how much there's available and how much people want it. So if the market works properly-"

"Markets working properly!" Frank laughed. "No-one wants markets to work properly. If they work properly you can't make any money out of them. Second rule of business, Igor."

Joe wasn't satisfied. "But even if they do work properly … I mean … what if they set the wrong limits? Are the limits they set always the limits which they should be?"

Frank snorted. "Course not, you daft git. Politicians'll never set the limits which there should be. Industry lobbies'll always make sure the limits aren't too tight.

"So … if the limits aren't right, then doesn't it mean that the price would be wrong?" asked Joe.

"Course it does. That's reality, mate," said Frank.

"So the price of carbon is always going to be too low anyway, isn't it?" said Joe.

The Professor nodded. "It is unless the politicians get very bold."

"Politicians getting bold!" chortled Frank. "That'll happen sometime around when we win the Premiership."

40. When money's no object

Pricing works as long as people are sensitive to prices of things, but they often aren't.

We made a long trip to Middlesbrough where the Clarets fought out a 2-2 draw. It was a freezing cold night, a gritty, bony, ankle-bashing game of long balls and cloddish tackles. It was northern England and inelegant but a point.

And then the next Thursday, like a breath of early spring,

Europe returned – glamour and glitter and anticipation sprinkled onto our dark industrial town – and we strolled down St James' Street with a confident, cosmopolitan swing. Paris Saint Germain were in town for the first leg and – well – we were about to hammer the French bastards.

We were on our way to the White Hart for a change. Joe asked where PSG were staying. Frank had heard from Spike that they weren't staying in Burnley at all. They were in the Radisson in Manchester. "Bloody frogs, we're not good enough for them, are we?"

The Professor clapped his hands. "You see – just what I was saying about prices. The fact that something is cheap doesn't mean people'll buy it. There must be a dozen hotels in Burnley cheaper than the Radisson in Manchester. And just because something is expensive doesn't mean people won't buy it."

"They probably don't even care what it costs," said Frank. "Just book a place, don't even check the price."

"If only our politicians knew this," sighed Igor. We'd entered the White Hart and found some seats while Igor brought a round. Joe tore open the pork scratchings and stuffed a handful into his mouth.

"A lot of policies to cut emissions are about exactly this," said Igor. "About pricing and making people pay for the carbon dioxide emissions they cause. The idea is that when the price of something goes up, people buy less of it."

"That's normal isn't it?"

"Except it doesn't work very well for gas or electricity, which are just the things you want people to buy less of."

"How's that?" asked Joe.

"To know that the electricity price has gone up, you'd have to read your electricity bill."

"Flipping heck, I haven't a clue what our electricity bill is." Joe turned to Doris.

"Don't look at me," she said. "It's direct debit. I haven't time to be checking the electricity bills all the time. Same for gas too."

"Waste of time anyway, you can't understand the bills even if you do read them."

"What?" said Frank in a serious tone, "You mean you don't know what a kilowatt hour is? Or a British thermal unit?"

Doris flushed. "How am I supposed to know that?"

Joe put his arm round her shoulder. "He's playing, love."

"Nobody knows what these things are," Igor consoled her. "Unless they're physics teachers or engineers. But let's imagine that you start noticing that you're spending more and more on electricity each month. What do you do?"

"Dunno." Joe shrugged his shoulders. "Probably just pay the bills and grumble about it."

"Let's imagine that you and Doris have been on a course and learnt about kilowatt hours and therms ..."

"Our imagination'll be hurting soon," said Frank.

"I dunno. Turn the lights off?"

"Let's imagine you've a list of all the appliances which use electricity and gas, and you know which ones use how much..."

"Ouch," said Frank.

Joe was exasperated. "How do I know what we'd do? How do I know what stuff to turn on and off all the time? How do I know which is the right heater to buy, which is the right bloody telly, the right bloody fridge, the right bloody kettle and the right bloody windows? And even if I did know somehow what to buy, where do I bloody get the time to go out and get it all? Who's going to come and fit all the stuff? Where do I get the money from?"

"What Joe's saying, Professor, is that just because the gas price goes up a bit, it doesn't make it any easier to afford a brand-new heating system and a new environmentally-friendly kitchen and bathroom," said Doris.

"Quite right," said Igor. "But something more. We do things because of habits and routine and the need for novelty and status and a host of other things going on in our mind. These are strong forces. We don't make decisions based on looking at prices and weighing up alternatives rationally, do we? Think what the price of petrol would have to be for you to cycle to work. Double? Triple?"

"He's right," said Frank. 'You'd have petrol riots in the street before you'd get Joe on his bike."

"But then it wouldn't be safe to cycle, would it?" said Darren who was wolfing down a sandwich before shooting off to training. "I mean if people were rioting, like."

Igor continued. "For a price to jerk us out of our habits it has to be much higher than the economist expects who thinks that we act based on rational decisions. And habits are one thing. Think of the people in Eastern Europe where they overheat their buildings terribly – 25 degrees or more. It's more than a habit. It's something deeply rooted in their culture and their upbringing. It's a reaction from the sufferings and privations of the past. Fear of catching a cold was passed on through generations, when firewood was scarce, when diseases were more dangerous, when they were refugees stranded in another country. Hiking the price of gas won't make them turn down the thermostat. If they have a thermostat, that is."

"This is … like psychological stuff again, isn't it?" said Joe.

"It is. It's deeply engrained in their psychology."

"Like I always say," said Frank. "You've to get inside your customer's head. That's basic business, chum. You can't change people's minds with taxes. You've to get inside their heads."

With that we put on our coats and hurried off to the match.

Price wasn't a consideration three hours later back in the Bridge after we'd defended heroically and survived with a 0-0 draw. We sat at the bar, jubilant. "Make it fizzy and make it French," shouted Frank, ever the master of schadenfreude.

41. Pricing and industry

Igor questions whether industry responds to carbon pricing.

"Out with it Igor … I can see you're dying to tell us."

The Professor sipped his Virgin Trains coffee patiently, but on his face there was the look of words trying to get out. "Not quite

Sant'Eustachio … Still… it's what people are ready to pay for."

"That's not what you were thinking, is it?" insisted Frank.

We were on our way to Birmingham again, for the Villa game.

So the Professor started. "Remember that the politicians want to make carbon dioxide emissions more expensive. It's not just for people like us – buyers of electricity and gas – but also for power producers and heavy industry. The idea is that if industry has to pay for its emissions, then it will take action to reduce them."

"Well that's logical."

"It's logical, but wrong."

"Doesn't that make it illogical?" asked Darren.

"It would be logical if heavy industry worked like other industries. It's logical where there are free markets and financial wizards; where you're free to stop producing something that's not worth it or where you can refuse to supply someone. Or where you can easily build new plants or shut down old ones. But that's not how heavy industry works."

"I know what you're getting at. It's politics and relationships," Frank nodded. "Just like the building trade."

"The men who run power plants are close to the men who make decisions about them. Perhaps not in places like England or Scandinavia, but in most of the world. They went to Karl Marx University together in Moscow or play in the same polo club in Buenos Aires. Or drink at the same watering hole in Delhi."

"Not like that in England? Come on Igor, it's rule three of business. It's not what you know, it's who you know."

"So if you see a threat on the horizon – like having to pay carbon taxes – the first thing you do is you start by doing nothing," Igor continued. "This is the tortoise industry. Things move slowly in the utility industry and most threats just blow away.

"Then if the clouds get darker, you start lobbying. Your local politician is more worried about the colliery band than about the water level on Vanuatu."

"Vanuatu?" asked Joe.

"It's an island in the Pacific."

"Paradise lost, eh, Igor," winked Frank.

"So the power plant managers and their political friends remind each other that theirs is only a small country. They can't make a difference. And they knock back their vodkas and order another round. Then if things get worse, they lobby the government and the government passes the parcel on to Brussels, and they win months or years as the bureaucrats go through the process of appeals and court cases and more appeals."

"So you're saying that making companies pay for emissions doesn't actually have much effect?"

"Well, in theory it can work. But you need free markets, you have to get rid of political meddling, you need unbendable regulators, you need strong financial and commercial management, it has to be easy for people to enter the market and get out of it..."

"And it's not like that anywhere?"

"In two or three countries in Europe it is. But I'd guess that 90% of the power plants in the world aren't like that.

"Then, if you really can't avoid the problem and you can't pass on the cost to your customers ... after a few years you might eventually start looking around for something to do about cutting your emissions. But what can you do if you run a modern coal-fired power station? Not a lot. You know you're out of a job so you just sit tight," said Igor.

Joe was mystified. "But it could take years for anything to happen, then. And they've all these schemes to make people pay for carbon emissions? And you're saying it doesn't make them actually cut their emissions?"

"Hmmm..." began the Professor. "It should work... but..."

"Come on, Igor, you can't be sitting on the fence," I said. "Does it work or doesn't it? You're the one who's supposed to know this."

"I think it does work, slowly. Over many years. Once the teething troubles are ironed out, once everyone is used to it, once the authorities know what they're doing... then it might work eventually, but it doesn't work quickly. And we need something to make them cut emissions quickly."

"So why do people in Brussels think that this works, if it doesn't?"

"I think it's because they spent a lot of time at their desks in front of computer screens and talking to each other. They don't get out much."

Frank smiled. "Heard that one before. Just like Man United fans. Watch every game on telly and they've never been to Old Trafford in their lives."

You couldn't have replaced Villa Park that afternoon with a 32-inch screen. It was as traditional, rough and tumble, end-to-end, mud and blood football as you could ask for. Villa had been in great goal-scoring form, but a flu bug and too many bookings had put Dunne and Cuellar, the stalwarts of their defence, out of action. The Clarets were at full strength. So were the fans, and the Park resounded with rival war chants and mutual mockery.

We went behind after eight minutes, Ashley Young drilling in a shot from the edge of the box when he was momentarily given too much space. We battled back into the game but couldn't find the break. They doubled their lead just before half-time … on the break from incessant Burnley attack, three against two … John Carew couldn't miss Milner's ball into the box.

But our grit paid off. We got level after only fifteen minutes into the second half. First McDonald found himself alone, unmarked, at the far end of the six yard box, thumping Blake's corner into the roof of the net. A few minutes later, a defensive mix-up on Villa's right let Elliott in, he danced his way into the box, cut inside and curled a peach into the bottom left corner of Brad Friedel's goal. Villa were dazed.

Then they pulled themselves together and began to press forward. We blocked and tackled and deflected and parried and tripped and shoved and barged and slid and dived … and then we used our hands. James Milner put Villa back in the lead from the spot. 3-2 to the hosts.

For Villa it was about a place in Europe. For us, survival. Heads down, focussed on the moment, we battled back, desperate to salvage a point. Time was ticking away, the eightieth minute had come and gone. A Fletcher header popped off the cross bar, and a scuffed shot by Paterson rolled past the post with Friedel wrong-footed. On 86 minutes we equalised − a glorious one-two with McCann put Fletcher one on one against Friedel. It was as if Friedel grew like a giant bat, whose ragged wings blocked the entire goal; Fletcher shrank to become his insect prey. But he was more man than we feared. A feign to the right and he slipped the ball left-footed over Friedel. It was 3-3.

I saw Joe glance over to Igor. His − I mean Igor's − face was taut, his hands clenched. Could it be that the tension was even getting to him?

Three priceless points were won at the death. Under the glow of the Villa Park floodlights, with rain crashing down and the stadium throbbing with the roars of dervishes and fanatics, we made one last push forward. Blake set Elliott scurrying down the right wing, he shot in a low cross and Paterson, a split second quicker than his marker, turned the ball in at the near post.

Amid the rowdy celebrations Frank remarked casually that the Professor seemed to have enjoyed himself.

"Oh, er quite ... yes, a very ... very interesting game," stammered Igor. "Quite extraordinary," he said, "how you take such pleasure in the ritual destruction of the enemy."

Frank winked to Joe and me. "I think you might be getting somewhere," he whispered. "He's in denial."

42. Signals

Why pricing signals are not as effective as policy-makers like to think.

We'd beaten Aston Villa away in the most exciting match of the century. Being in the relegation zone didn't matter for now. Whatever a gauntlet was, we'd flung it down.

The train was stuck somewhere between Stafford and Crewe. The inspector said it was signal failure.

"The reliability of signals …" began the Professor.

Joe and Frank groaned. Doris made herself busy getting out sandwiches and the tea flasks. Kelly was plugged in to her iPod and fast asleep. But Darren was listening.

"Go on Professor," he urged. "It's more interesting than Uncle Frank's property games."

"Another reason why people argue for putting a price on emissions is for the price to act as a signal."

"What do you mean a signal?"

"It's a signal to investors about what to expect. Imagine … let's say Zbigniew in Poland. He's planning to build a new power station. He's making a choice between 1,000 MW of coal or 1,000 MW of natural gas. This monster is going to run for 40 years.

"If he builds a coal-fired plant, that's over 200 million tons of carbon dioxide over the forty years. If he builds a gas-fired plant, that's about 100 million tons over the same period. Now, why does it matter to him? Because he has to pay for the carbon dioxide emissions. If one ton costs 20 Euros, then he has to pay 4 billion Euros for his emissions with the coal plant and only 2 billion for the emissions from the gas plant. So with the gas he'd save 2 billion Euros! Normally, the coal plant would be much more profitable – he's sitting on vast seams of good Polish coal. But if he wants to burn gas he has to bring it in from his old friend Yuri in Moscow."

"So how does he decide which to do?" asked Joe.

"Well, he factors in the cost of purchasing emission allowances. If the price of allowances is low, it's probably still worth building the coal plant. But if the price is high – say 25 Euros or more – then it's probably better to build the gas plant."

"But how does he know what the price of carbon allowances is going to be? I though you said it goes up and down all the time," said Joe.

"He has to make a prediction," said Igor. "He makes a

prediction based on the signals given by the government. If he believes the politicians that they'll keep the carbon price high – you know, if he really believes them – then he'll go for the gas-fired plant.

"What if he doesn't?" asked Joe.

"Believe the politicians?" exploded Frank. "Was he born yesterday? Believe them self-important unelected bureaucrats in Brussels? Believe them little farts in Westminster with their la-di-dah dinners and duck ponds? Bollocks!"

The Professor blinked at Frank's outburst. "That," he said, "is the problem. If he doesn't believe they have the stomach for high carbon prices … then he'll go for the coal plant. And of course, once he's built the coal plant, there's no way back. Unless you want to bulldoze it. It's there. It's stuck there for another forty years."

"So it all comes down to whether you believe in what the politicians are saying?" said Joe.

"It does. He is only going to believe the politicians if they're credible, authoritative, consistent, and … un-bending. In practice, like railway signals, price signals go up and down and up and down. Sometime the signals work, sometimes they don't. It all depends on how the political winds are blowing."

"Not much hope then, really."

"Well, let me put it this way," replied Igor. "If politicians were railway signalmen, we wouldn't use trains."

43. False choice

The theoretical choice in emission trading between paying for your emissions or actually reducing them is not a real choice.

It was the Wednesday evening before we set off to Paris for the PSG second leg. Darren was in ecstasy. He'd just got a call from the youth team inviting him over for a trial.

"That's the real way to build a team you know, Prof," said Joe.

"It's not just about buying stars from Brazil for twenty million quid and that. Get 'em in young. Costs nowt compared with doing big transfers. Course, it takes time. You can't build success at the youth level overnight."

"No-one can wait these days," said Doris. "Everyone's in a hurry. They buy success like they're down at Tescos. It's not football any more."

"Yeah, right. Look at Real Madrid. Cost £350 million. Bloody joke. All bleeding prima donnas, anyway. Same with Chelsea," said Frank. I winced at mention of Chelsea. "Sorry, mate," said Frank. There were some things even Frank could sympathise about.

"If I understand correctly, Arsenal take a different approach, no?" said the Professor thoughtfully. He was in his favourite armchair by the gas fire and was sipping a very nice wine he'd brought round for us all.

"Yeah, there's a few like that … but most just go into the market and buy players," explained Joe.

"Well, you know it's just the same in industry," began Igor. He swirled his glass, waiting for a cue.

"How's that?" asked Joe, obliging.

Igor put down his glass and sat back. "I think I once explained to you this emission trading scheme which they have in the European Union."

Frank snorted. "Yeah, we've heard all about that. It's that scam where they're like buying and selling air. Some big tax dodge, or summat. Chopper had a mate doing it. Got nicked."

The Professor laughed. There had been problems with emission trading, but … Joe had forgotten how it works so the Professor decided to explain again. The European Commission sets a cap on how many emissions Europe's industrial companies can make in each year. Say 1.6 billion tons a year. European governments issue to industrial companies that many "EU allowances". Each allowance gives the right for a factory to emit one ton of carbon dioxide – it's a kind of permit. All industrial companies – power stations, cement plants, steel mills and so forth, about 14,000 industrial facilities across the EU – have to report their emissions of carbon dioxide each year. And each year, for each ton of carbon dioxide they emitted in that year, they have to hand in to the authorities one emission allowance.

"But how does that make them cut emissions?" asked Joe.

"It makes them cut emissions because the total number of allowances issued is less than the current level of emissions from all the factories. So somewhere, there will be a squeeze and some companies will have to cut emissions."

"Do they have to pay for the allowances?" asked Joe.

"In the future they will," said Igor. "There will be auctions. But today companies get most of their allowances free from the government at the beginning of the year. But they still don't get as many as they'd need."

"Freebie from the government," interrupted Frank. Sounds like my kind of game, that, Igor."

Doris told Frank to stop butting in and let the Professor say what he has to say.

"So a company in the scheme… Let's take the cement plant in Clitheroe just down the road from here. Let's say its emissions are 500,000 tons of carbon dioxide in the year. Perhaps they get from the government 400,000 allowances."

"So they have 100,000 fewer allowances than their emissions," said Darren.

Frank butted in. "Obvious, isn't it? It just goes into the market and buys 100,000 more allowances, doesn't it?"

"Not so obvious, my friend. One choice, it's true, is it can buy allowances. The other choice is to reduce its emissions. If it cuts its emissions to 400,000 tons it doesn't need to buy any more allowances. You see?"

Joe sat up. "You're right, it's just like Burnley… To strengthen the squad we can either buy someone on the market … or we can bring lads like Darren up through the youth team. One takes a few weeks and costs a lot … the other takes a year but might not cost so much."

"A good few years and even then it might not work," added Frank. "You take in a dozen lads … only one or two'll make the grade, if you're lucky."

"It's exactly the same choice for our local cement plant in the emission trading scheme. It can either buy in carbon allowances or it can do something to reduce the emissions. To buy carbon allowances is instant and painless. In fact, it's exciting trading on an exchange with someone else's money.

"And think of the alternative – actually to reduce emissions. Nothing but hard labour. Five years or more of stress and sleepless nights to make an emission reduction project happen. You're dealing with managers and boards of directors and employees and suppliers and engineers and authorities and local politicians and lobbyists and people who want to receive kickbacks and then people who want to give kickbacks, and … the list is endless."

Doris had been ironing and making sandwiches and packing bags. Now she came and sat down on the sofa next to Joe. "Haven't you lads had enough of this?" she asked, yawning.

"We're almost done, my dear. My point is that to do a project to cut your emissions is a terribly big thing. It's hard work, takes many years and the man who starts the project is probably not even working for the company by the time the project is finished. It won't be his glory if the project works. But if it fails, he'll surely be the scapegoat."

"Then who'd bother reducing emissions – if it's such a hassle and it's way easier to buy the stuff?" said Joe.

"And remember how uncertain it is," said Igor. "When you start building your project you have to assume that the carbon price will be such and such – and by the time you have got to the end of the project, the carbon price could just collapse into smoke and the project wouldn't be worthwhile anymore. The price swings up and down all the time, you can never be sure that your project is going to be successful."

"So it's not a real choice at all, is it?" said Joe. "I mean, the choice between buying carbon allowances or cutting your emissions. No-one in their right mind is going to bother with cutting emissions."

Doris yawned again and stretched, reminding us why the heating wasn't on 15 degrees. "Now it really is time for bed, boys. We're up at the crack of dawn tomorrow and we've a train to catch to Paris."

44. Le prix d'une bière

Having one carbon price in a market is good if you want to minimise the cost of cutting emissions, but not if you want to cut emissions as quickly as possible.

Once we got to Paris, Joe and Frank wanted to walk from Gare du Nord to the Parc des Princes. Together with a dozen other Claret faithfuls.

We popped into a Café-Brasserie on Rue Maubeuge and knocked back a couple of glasses of vin de table.

"It's his fault," said Joe to Doris, gesturing towards the Professor. "He started this wine lark."

We were gasping by the time we got to Boulevard Haussman. A potential incident with a barman in a PSG shirt dissolved into bonhomie and another round and a toast to Parisian women.

Then Joe was almost knocked down by a 2CV crossing the Champs Elysées and Doris said it served him right, because she was still sore about the Parisian women, which led to Joe spending more than he could afford in the shops on Avenue Montaigne.

Chastened by the tiff, we didn't stop again – except for taking photos of the Eiffel Tower and a Ferrari store Joe spotted – until we found a bar at the end of Rue Molitor. Kelly was fed up and very thirsty and Doris needed a cup of tea.

Joe stood at the bar and ordered a beer. As he moved back to join us, the barman called to him angrily, gesticulating like a Frenchman. He was a Frenchman. Igor jumped up and began to interpret. He explained to Joe. "There's one price for drinking at the bar, and there's another price for sitting down. You paid the bar price, and now you're coming to sit down. You can't do that."

"You're joking!" said Joe.

"It's true."

Frank roared with laughter. He'd never heard owt so bloody daft in his life.

Igor said that he didn't think it was daft. It was quite common to have different prices for the same thing.

"No it's not," said Frank. "It's daft."

"What about match tickets, then?" said Joe, challenging his brother. "Match tickets are cheaper for pensioners and kids, aren't they? And if you're stuck at the back."

"You see," said Igor. "It's not so unusual. There's just the same problem for these emissions trading markets."

"You've not found another problem with emission trading here in a bar at the Parc des Princes," laughed Joe, forgetting the upset.

"Oh yes I have," replied Igor. "You see-".

"Hold on, Igor," said Frank. "We're a bit peckish. Can't you order something to eat first."

After a flurry of translating and careful of elimination of molluscs, amphibians and working animals, we settled on braised rabbit and Igor picked up the trail of his thoughts again. He explained that a lot of people believe unquestioningly in the power of markets to get industry to cut emissions. The trouble with markets is that by definition markets have one price. The same price for power companies as for cement companies and steel companies and brick factories and so forth."

"In reality, different industries can afford different prices for carbon dioxide emissions. A coal-fired power plant would be ready to pay no more than, say 15 Euros per ton of emissions. Anything more than that, it makes more sense to switch over to gas. A cement manufacturer may be able to pay up to 50 Euros for emissions. A car driver would be ready to pay the equivalent of a couple of hundred Euro per ton of emissions, before he decides to take the bus.

"A single price," continued Igor, "either kills you or leaves you untouched. It's a badly calibrated lawnmower. If you're in a big hurry to cut emissions, then you need all hands to the pump. You need everyone to want to cut their emissions. A single price won't do that."

"Right," said Joe. "Just like it's no use having a single price for match tickets. Either you'd not get any kids and oldies in, or you'd have a massive fight at the turnstiles because everyone would be there."

45. The madness of long-termism

Emissions markets might be a good long-term solution, but as we are talking about a short-term problem (to stabilise emissions), we should be looking at short-term measures.

Joe and Frank went off to check where our turnstile was and the rest of us waited in the bistro. A few minutes later the brothers were back. "They're not letting us in for another hour," said Joe. "This froggy copper told us to bog off and come back later. I think we'd better have another ale."

Doris spoke up. "You aren't having another ale, Joe Sugden. You've had enough ale to sink a ship today, and enough wine to sink another. That's enough. You'll have a headache like a pickaxe's gone through it in the morning. And you'll stink," she added.

She turned to the Professor for support. "It's always the same. They get pissed now, and completely forget about tomorrow."

Igor sympathised. "All today, no tomorrow − it's the scourge of society."

"Well I believe in short-term," said Frank. "Long-term is just a lot of short-terms stuck together, end to end. What's the difference? If you win every match, you win the league. It's simple."

"Except …" continued Igor, "except in the case of climate policy. In climate change policy short-term thinking is now needed more than ever. Desperately so."

"Short-term thinking?" said Joe. "That's a new one."

"I must have told you before. We've two or three years to stabilise emissions. To limit warming to two degrees we can emit no more than 400 billion tons of carbon dioxide and other greenhouse gases from now on. Not until 2050 or until 2100. Just full stop. We emitted 40 billion tons in 2008 and that's growing. At this rate we will have run out of our budget by 2020. The clock is ticking very loudly. Very loudly…"

We were quiet. Igor continued. "It's an annoying old alarm clock with bells on top. The wake-up will be … horrible.

"This means we have to stop any growth in emissions by 2011 and after that we have to cut them by ten percent a year. If we wait much longer, we risk getting to the point where we can't avoid breaching the limits."

"We have to focus ruthlessly on the immediate present. It's the only way we can make it. 2050 and 2020 are completely irrelevant. We should not be worrying about them. For winning time against climate change 2050 and 2020 are just a distraction."

"You mean all these international conferences and stuff are – "

"Yes," nodded Igor sadly. "Little ants on autopilot…"

"Just wasting taxpayer's money, that's all," said Frank eagerly. "If they worked for me they'd be focussed on the next couple of years and that's it."

"Well why aren't they, Prof?" asked Joe.

"You ask me very … sensitive questions, my friend. Where to start? Partly it's because politicians, with few exceptions, find it comfortable making bold plans for a time when they will be out of office or propping up a zimmer frame."

"Wasting taxpayers' money, like I said," said Frank.

"It's partly because academics, with some exceptions, prefer to amuse themselves with elegant long-term models instead of tackling the really tough gritty, smelly, messy, bruising problem of what to do tomorrow…"

"Lazy buggers," said Frank. "That's all taxpayers' money, too."

"It's partly because democracy has hit a brick wall … but liberals can't contemplate such a momentous admission."

"Tossers," added Frank.

"And partly because the greedy right wants to squeeze that last bit of oil out of the ground before … before it has to confess that … it was wrong about the oil."

"Well … er," said Frank.

"We can't reconcile savage focus on the short term with the habit of rejecting short-term thinking as the problem."

"Exactly," said Frank.

"Our policies should be desperately focussed on steps to cut emissions drastically in 2010 and 2011 and the next years. The long-term is a distraction from the real business of cutting emissions."

46. Praying

"Rational" policy-makers, economists and scientists have their little religions just like the rest of us.

After ninety minutes it was still nil-nil. How, we'll never know. PSG attacked incessantly, but we defended stoutly at times, desperately at others, on occasion outrageously flukily and sometimes despite ourselves. God proved to be an Englishman once again, the woodwork coming to our aid on three occasions. The French just couldn't get the ball in the net.

Extra time came.

"Just one goal," prayed Joe looking to the stars. But it was a cloudy night, and no goal came.

Extra time passed and now it was penalties.

Joe couldn't look. Frank clenched his fists.

Jordan missed the first. But Makalele blasted his effort into the crowd.

We were deep in prayer. Praying for all things Claret. Praying for Jimmy MacIlroy. Praying for Billy Hamilton. Praying for Turf Moor.

Then Coupet was sent the wrong way. So was Jensen.

Alexander scored. Erding scored. So did Fletcher – lucky, in off the post, but it counted.

When Clément missed even the Profesor whooped for joy.

"Just score, just score" breathed Joe to McCann a hundred yards away. "Just score."

He hit the post.

"Please miss, please miss," breathed Joe to Giuly.

Jensen flung himself to his right, parried the ball, and it spun away to safety.

We were through. All were jubilant. It was a moment of glory. A smile had broken across the Professor's face.

"There's someone up there," said Doris afterwards in the hotel bar. She'd seen Darren and Kelly to bed and returned to join us.

"And he's got a claret and blue scarf," said Joe.

"The power of faith," murmured Igor. "It's all down to faith.

He turned to Joe. "You're not alone in having unshakeable faith in Burnley. Many policy-makers have a similarly deep faith."

"What, you mean they're Burnley supporters?" said Joe.

"No, not faith in Burnley. Faith that when we reach a river, there'll be a bridge there. It's like Darren's game Age of Empires – as you explore new areas the grass and trees and paths magically place firm ground before us just before we fall into nothingness."

No-one got it. "People who believe in the science and the logic of the market tend to look down on faith and mysticism. But in fact they have their faith just like people who live in forests. Faith that we can replace the infrastructure around the world in a few decades. Faith that engineers will float the ark of carbon capture and storage in time save us. How many miracles will that require?

"Faith," Igor continued, "that industry and consumers will respond alertly and briskly to their faltering, flickering price signals. Faith in the market. A bit shaky, but still there.

"Faith that one plan will be enough.

"They're setting targets for 2020 and 2050 even though they know that emissions must stop growing by 2011 and be cut dramatically immediately after that. They hope blindly that we'll be OK.

"Perhaps in some places things will work to plan. Where instructions from the centre are executed efficiently and with no emotion. Where there are no ladies with banners and handbags holding up the permitting process. Where a technological dream is not a myth. Everywhere else … all they can do is pray for rain."

The Professor had his speech and felt a lot better for it. But we were too excited and too tired to follow. We ordered another round and turned to watch the highlights on French TV.

Part III
March to May

The Professor's Burnley Protocol aims to save the planet if all else fails, but can the boys save Burnley in the Premiership?

47.   A night on Pendle Hill

Igor reminds us that we are up the creek.

"What should we do?" Joe asked Igor. Our walk in the country had not gone well. It was raining heavily, fog was descending and it was late. We were stuck on the top of Pendle Hill.

"You ask the right question. But I don't think that you've yet grasped how serious the problem is. We started by looking at some causes of the problem. We then considered the ways in which the elected and non-elected representatives are trying to deal with the problem."

"I mean about getting off this bloody hill," said Joe curtly.

"Gravity, my friend. You can rely on the laws of physics for the foreseeable future. Eat your Mars bar and calm down.

"I was referring to the seriousness of our predicament on the planet. I don't think that you or Doris or Frank understand it fully. I don't. I have erred in not discussing it enough with you. I assumed you understood the horror. But you don't. You don't feel revulsion at the destruction of coral reefs, the loss of glaciers, the nudity of Kilimanjaro, the disappearance of the Amazon forests, the peat fires sweeping Indonesia, the suffering of gorillas and orang-utans, the spread of desert, the salination of low-lying islands, the death of albatrosses and the acidification of the oceans, the murder of whales, the destruction of mountains in Kentucky by coalminers, the weariness of polar bears, the desiccation of tropical islands, the bloodshed when cities collapse, the random violence of angry, hungry, fearful people … My friends, darkness is drawing in around us."

"I know, Prof. But if we don't get off this mountain, we'll have to spend the night on it," said Joe.

"This is no time for joking, Joe. I meant that darkness is drawing in around life on this planet. And it's not human beings who'll be the survivors, but cockroaches and ants and scorpions.

"Everything we do touches nature and harms it. There are

seven billion of us on the planet destroying the systems we depend on: the cycles of oxygen, nitrogen, water, interactions of the oceans, the land and the sky, and the agents of these interactions, millions of living species of plant and animal in delicate and complex balance. Now climate change is bringing irreversible changes to these systems and we don't know where it's taking us."

Igor stopped to get his breath. He found a rock and sat down on it. He sagged under his rucksack and let out a deep sigh.

"Without nature there's nothing. You could try building some system with steel pipes and concrete and tubes and mirrors. But what's the chance that it would work? 50%? Is that a good bet when all life is at stake? We don't understand nature enough to make it our slave."

"But, Igor ..." protested Joe.

"Exactly, my friend, it is scary. And we are trapped."

It was now virtually night. We could hear a beck gurgling somewhere below us. The occasional grunt of a woodcock. Joe found a dry-stone wall and we felt our way along it. Then it petered out and Joe tripped over a fallen stone. "Bugger, I've done my ankle."

I could just make out Joe in the gloom, thrashing about in the bracken. I helped him hop over to where Igor had set up a temporary base. We conferred and Igor decided we should stay on the mountain tonight. He thought we'd be fine if we wrapped up warm and had some smokes and chocolate. He gave short shrift when Joe protested that the Professor was far too old to be out on a mountain at night.

"Old? Me old? You're mistaken, my friend. Old in body, but not in spirit. I've survived two wars and five wives and I'm still in fighting form. Wrap up well, and hurry up with those cigarettes."

The three of us sat down by the stone wall, out of the worst of the rain and tried to light up our damp cigarettes.

"As I was saying the situation is very grave. Fundamental problems which don't go away, even when very clever people deny them.

"We have many people on the earth. And the number's getting bigger quickly. But it's political suicide to mention population. Many of those people consume far more than nature can provide. And as they move from country to city, they consume more and more.

"And democracy has failed us completely. Democracy and peace."

"Peace?" I exclaimed. "What do you mean?" I sensed that the Professor was sharpening his knives again and I didn't like it.

"Democracy is the relationship between government and citizen, isn't it? It's what gives the citizen his say, his freedom. But we know that most citizens are incapable of voluntary restraint – "

"Voluntary restraint?" asked Joe.

"He means – " I began.

"Holding back," the Professor corrected himself. "Joe, the way we live is damaging ... destroying the natural systems we and other species rely on for life. To stop that we need to hold back a bit ... take it easy. But most people don't think beyond themselves. In every decision, whether shopping or voting or resting or eating, our preference is for ourselves and the next ten minutes. Nothing more. So democracy is incompatible with the behaviour needed to protect the life around us. Unless there's a very strong culture which counterbalances our instincts."

"We understand that, Igor. But you said 'peace'," I remarked.

The rain was falling harder and thunder rumbled somewhere beyond Clitheroe.

"In the way that democracy defines the relationship between government and citizen, peace defines the relationship between government and government."

"You what?"

"What I mean is that if Britain wants Japan to stop killing whales, for example, it will negotiate with Japan and at most it might one day impose economic sanctions or ban Japanese leaders from coming to Britain. But it'll never go to war over whales."

"Why not?" asked Joe. "It would teach them a lesson." Meanwhile he'd taken his left sock off. His ankle didn't look good at all.

"Because we've learnt that war is so horrible, my friend, that we do everything to avoid it. The desire for peace overrides any principles we have about whales or forests or emissions of greenhouse gases. It wasn't always like that, but now that weapons have become very powerful..."

"So all we can do is talk," I said.

"And talking is very slow. Especially when you recall how people talk. In international negotiations we have the talk of bureaucrats and lawyers and – bumbling idiots who can't see the wood for the trees. It takes days to change small sentences, weeks to change a paragraph. And decades to change anything of substance. That is the failure of peace. With greenhouse gases there's not so much time."

"Are you saying that Frank and I need to join the territorials?"

Igor paused. "The war was bad. Very bad. I don't know which is worse."

We sat miserably in the rain. A long cold night lay ahead of us. Joe lit up another cigarette.

"Cheer up, Igor," said Joe. "Things can't be that bad. We're out of the FA Cup, we're out of the Carling Cup, and we're 18th in the league and we just lost three-nil at home to Portsmouth … but we're still in Europe and we're off to London tomorrow. Things could be worse."

But the Professor was fast asleep.

48. Joe's dream

Joe realises that the future might not be so fun.

At dawn the sheep woke us. "Oi," shouted Joe. "Bugger off!" He jumped up to shoo them, and then remembered his ankle. He screamed and sat back on the grass. We hobbled down Pendle Hill that morning. Me supporting Joe who hopped and squelched through the marsh like a bathroom plunger. Igor looking haggard and sorry for himself and constantly sneezing.

"Survival skills," he kept muttering. "Don't forget this. You may need them if it all goes wrong."

After an hour we spied civilisation and knocked on the door at the back of the Assheton Arms. The cook soon had us tucking in to bacon and eggs and toast and baked beans and piping hot tea. Meanwhile Igor had found his torch, his map, and his mobile phone at the bottom of his rucksack. Once Joe had called Doris he gave the phone back to Igor: "It's OK. We're clear to go home."

In the van back to Burnley, Igor dozed off. Joe was very quiet. "You all right, Joe?" I asked. "Not something Doris said? Did you get an earful for staying out the night?"

"No, nowt like that."

He drove on in silence.

"Come on, mate, there's something up, isn't there," I said.

"All right then," said Joe. "I had a dream last night. It were weird."

"Go on."

"Well .. sounds stupid, but … it were like this…."

"It were early evening and I was sitting having a beer in the living room. Watching the FA cup final. Burnley were in it and losing 4-0. Doris was in the kitchen making tea. I can remember … she was doing pasta and tomato sauce. Like we had in Italy. I'd even fetched some garlic from the garden. It was dead hot outside. So I was just sitting there watching the match when someone knocked on the door…"

In his dream Joe sighs, takes a swig of his home-brew and goes to the door.

A Bangladeshi chap is standing there with a Tesco bag containing all his belongings.

"Right, mate?" says Joe.

"Greetings," he spoke in the Queen's English. "Can I please have somewhere to live. I was flooded out. I have just arrived from Bangladesh."

"Sure," says Joe. "No problem. We've got a spare room in the attic. Tea is in ten minutes. Get yourself a quick shower and we can get a beer in before."

Joe went back to the match. A few minutes later there was a knock on the door.

"Hi," it was another Bangladeshi chap. "I need somewhere to live. We were flooded out."

"Ok then," said Joe. "But you'll have to share with Rashid."

The new arrival turned and beckoned to his family who were hiding behind the garden fence. As it was a dream, they all fitted into the spare bedroom.

Then they were sitting down to supper, sixteen of them around the table, all squeezed into the front room, pasta and sauce everywhere all over the floor and on the walls, babies crying and toddlers trashing the dishes. Cans of beer piled up everywhere, carpet was soaking. Joe was still trying to watch the match amid the chaos. Then they heard a vehicle pulling up outside. Doris got up and peeked through the net curtains and saw a bus parked in front of the house with a Chittagong number plate…

"It was so real. I just thought to myself, where are they all going to go? How are we going to feed them all? There were hundreds of them forcing their way into the house. Darren went mental and Doris was screaming and Kelly was crying, then I realised that it was them sheep bleating in the next field."

49. The first years

Doris figures out that a clue might lie in the first few years of our lives.

All trains to London on the Sunday were cancelled because severe storms across the Midlands had taken down the lines and there were no replacement locomotives available. There was no way to get to the Arsenal match. It was the first league game Joe had missed for years.

Igor was relieved not to have to travel to London and back straight after coming off Pendle Hill. He went off up to bed with a hot whisky and lemon. I went over to Joe's.

While we watched the Clarets sink to a 5-1 drubbing at the Emirates Stadium, Doris was doing some housework. Joe had his leg up on the coffee table and a bag of frozen peas tied round the ankle and a can of Stella. He was pondering about Igor: "I don't get him at all. I mean he's perfectly normal, isn't he? He likes birds, doesn't he? How come he doesn't like football, then? It's not like he's, like, you know."

Doris tugged one of Joe's shirts out of the pile of clothes and shook it straight and laid it on the ironing board. "Of course, not, love. He's fine. And you can't say he's not trying hard, can you?"

"He's done well on his facts and figures. You can't beat him there. But … like last week in Paris when Giuly missed and we won. He smiled. Smiled! I mean it's like …"

Doris turned the shirt over and pressed its back.

"It were like we were all screaming and shouting and Frank was wild and you were like some bloody mad thing, and I look over to him, and he's standing there like a daft twat smiling." Joe pinged his can into the waste bin and reached over for another. But he was leaning too far back and couldn't reach it. "Love?" he pleaded to Doris. She sighed, put down his folded shirt, and handed him the fresh can. "Love you," he said.

Doris went back to the ironing. "Well he's not from here is he? He didn't get it when he was a kid."

Joe looked over to Doris curiously. "Course he didn't get it when he was a kid. It wouldn't be right, would it?"

Doris put down the iron. "I don't mean it, you loony. I mean, he didn't get football. His mum was a foreigner. He probably went fishing or whatever they do in Russia. It you don't get it when you're little, you never will." She shook out the next shirt. "And you know what, I reckon it's the same with his stuff. If you don't get it drummed into you when you're little, you never will."

"You mean trees and stuff?"

"Yeah. And carbon stuff. I mean why do the kids always want new things and why do they want to go places all the time? We didn't did we? It's us, isn't it?"

"Like, Frank and me support Burnley because Dad did? From day one."

"Exactly," said Doris.

"And that's why he doesn't get it," said Joe.

"Exactly."

"Don't start that bloody exactly, you sound just like him, you do."

50. Vision

The idea of having a vision for the future is rubbished by Frank.

We gathered again before the game against Zenit St Petersburg. It was a home leg – they were among the first Russians ever to have been to Burnley.

"We're going to win the Europa League!" sang Joe as we cracked open some cans.

"Aren't you celebrating a little early, my friend?" asked Igor. He'd been sitting hunched over his notebook, chewing a pencil, drinking hot tea and snuffling.

Frank arrived. "How's your plan going, Igor?"

"I am making good progress, thank you. I have determined that the next thing we need is ... a vision of the future ... where we want to get to ..."

"A vision?" asked Joe.

"Yes," said the Prof. "Like ... Burnley winning the Premiership in 2015."

"Not that bullshit, Igor," said Frank.

The Professor was taken aback.

"Vision won't work," continued Frank. "You can't eat vision, you know. You can't wear vision. People on the street don't want vision. They want stuff to do. They need to know what to do now and tomorrow not in ten or 20 years."

"But, how can we plan for the future without vision?" Igor asked.

"S.A.F."

"Sorry?" Igor was puzzled.

"Sir Alex Ferguson. He's the manager of a team down in Manchester. 'We aren't thinking about winning five trophies we're just thinking about winning the next game.' Vision is your next game."

"Is that what he said?" asked Igor.

"Something like that. He doesn't dream about glory or records. He just focuses doggedly on the next game."

"Interesting …" mused the Professor.

"Look at us. We're in the shit this season. 19th place. We need a good eleven points from the last nine games to get out of the shit. We don't need any bleeding vision. We need to win the game against Wolves on Sunday. That's it. And then we need to beat Wigan. And then Blackburn. Game by game. And I don't care if we win by an own-goal in the 93rd minute or are ten-nil up at half time. We just have to win the next game.

"Igor! Listen to me for once. Get your bloody nose out of your book. Vision kills. It distracts you from doing things now. Vision is for bureaucrats. For dreamers. For idle bastards who work in offices and have nowt better to do. Vision is an effing cop-out. It's easy to create a vision. Give Kelly a pack of felt-tip pens and she'll draw you a vision in ten minutes. If you want to save your bleeding penguins, then you need action, not vision. You need to get your arse into gear. The best thing to do would be line all the visionaries up against the wall …"

"Please, no. I've been there." The Professor shuddered. It was a shudder of dark memories.

Frank shrugged. "Whatever. You know what I mean."

"I do. And I must say-"

"Igor. Utopia's bullshit. It's dangerous, Igor. You know about history. Folk don't like utopia. It's been promised to them before. By bad people. Let's just go step by step but be brisk about it."

"Then we shall have no vision," said Igor.

"You shouldn't care what the future's like. You just want to dig us out of the shite now."

"We shall always be focussed on the next game, my friend," said the Professor. He knew when he was beaten.

51. The Burnley Protocol

Igor introduces his plan.

The Zenit game? We put our heads down and charged. It was as bloody as Crimea. At the end of the first half we took the lead – against the run of play – when Eagles was upended in the box. He picked himself up and swept the penalty past Malafeev in goal. We doubled our lead half way through the second half. Paterson showed quick wits and took advantage of a poor pass back by Yurianov. Then we pulled down the shutters. Zenit attacked without pause in the last fifteen minutes and got their away goal – a glorious curling effort from the right edge of the box. There was nothing Jensen could do about it. We had the edge, they had the away goal.

"You see," said Frank as we spilled out of the ground afterwards. "It's just about focussing on each game."

"Good, good. As I said, we shall have no vision," muttered the Professor.

By the time we got to the Bridge it was almost time for last orders. Joe tottered back from the bar carrying pint jugs like a Bavarian wench, but everyone was so tired they could barely manage it. The Professor took a sip of his pint. "Then I would like to propose the Burnley Protocol. The Burnley Protocol is how we are going to tame the problem."

"Burnley?" Joe raised his eyebrows.

"Yes, I said Burnley," replied Igor. "We want something practical and down to earth and short-term, and closer to home. The Rio Convention, the Kyoto Protocol and now Cancun – they all sound very exotic but bring us nothing. They're just too far away.

"The Burnley Protocol is a backstop. It's there in case we can't replace all the power stations quickly enough; in case carbon capture doesn't work in time; in case trading schemes and taxes don't have the effect we hope. The Protocol contains three deals. A deal with nature. A deal with the rich. And – "

"You what?" Joe interrupted the Professor.

"Let me explain. The deal with nature … we have to learn to live in harmony with nature. Once the deal's in place our ambitions will be much less harmful … we simply won't need to use so much energy … so the job of the technologists and those replacing our infrastructure will be much easier."

"Pie in the sky," said Frank.

"Now, the deal with the rich," continued Igor. "The rich influence how we live. Now they need to influence us to live the right way."

"Hogwash," said Frank.

"And finally the deal with …" began the Professor. Then he looked up. Frank, from the smirk on his face, was texting to Chopper or Baz. Joe had dozed off and emptied his crisps over his trousers. Doris had nestled onto his shoulder and her eyelids were drooping. "Another day," I said.

52. A bond with nature

Why we need to cut a deal with nature to avoid wrecking the systems we depend on.

Igor spent the whole of the next day tidying up his papers. Several bags of waste paper had collected, which he made me take to the recycling – on foot. Which was fortunate, because just at the recycling centre I bumped into a mate who said that the he'd heard that the Chelsea fan had returned to London and it was all off! Fantastic! If I played my cards right I'd soon have my own bond with nature and not what Igor was thinking.

On the Saturday afternoon, the day before the Blackburn game we went to Joe and Doris' for high tea. The table heaved with Scotch eggs, tongue, gammon, beetroots and gherkins, salmon sandwiches and cheese sandwiches. Jam rolls, chocolate hobnobs, scones, a Victoria sponge, a Battenberg cake and fresh strawberries came later. Crisps. Pots of tea. Cans of beer. Orange juice. A feast.

"Feed a cold," said Doris to Igor. He was still coughing but his appetite was robust. He'd already chomped through half a dozen slices of bread with cheese and pickle. "No thank you, my dear," he raised his hand as she offered him the plate of gammon. "Delicious, I know, but it's not my meat day."

Once he was done, he leaned back and dabbed his brow with his handkerchief. "Now you can tell us about your deal with nature," said Doris.

So he did.

Except for occasional violence on the terraces, our societies are peaceful because there's a deal between man and society brokered by the rulers. The deal sets out how we live in society and it limits what individuals may do. It keeps people in balance among themselves and holds leaders back from tyranny and people back from anarchy.

Frank snapped that this was all perfectly obvious to anyone with the slightest intelligence.

"You're underestimating yourself, Frank," replied Igor.

Doris laughed. "Go on, Prof. We're listening."

"We have a deal among men, but we don't have a deal between man and nature. And we need one because whether you have communism or fascism, liberal democracy or a libertarian paradise, human nature will destroy the planet unless there's a contract with nature in place."

"But it's a bloody daft idea," protested Frank. "How can you have a deal with nature? You can't negotiate a contract with a bleeding earwig. Even Winston only understands basic instructions. Not enough to enter into a contract with him."

"Forget Winston" retorted Igor. "He is practically human anyway. Now, be serious."

Frank said that things will sort themselves out. The Professor smiled. "You have no reason for optimism, Frank." He said that nature's big problem is that she doesn't warn us when we are overstepping the limits. Emissions are caused by the Americans and the Europeans, but it's the Bolivian and Kyrgyz mountain people who get it in the neck today. Innocent as canaries in a mine. Nature's systems are flawed because they don't give it back to the people who deserve it. So we have to make this good by way of our contract.

"There's only one planet. We don't have a separate ground to train on. We can't just roll out new turf every season. Nature's been doing her side of the deal for hundreds of thousands of years without complaining. But now she's about to crack."

"So what's the deal?" asked Joe.

"The deal is very simple: we establish a deep bond with nature. That puts everything in perspective and we stop wanting to do all the things which damage the planet."

"A bond with nature? But Professor, that's just … dreaming, isn't it?" protested Joe.

"For goodness sake, love," said Doris. "Let him talk. It's interesting this. It might actually help us."

Igor turned to Joe. "There are people who already have a deal with nature. The Shuar Indians of the Amazon forests of Ecuador and Peru. Have you heard of them?"

Joe hadn't. "You mean Indians … like … spears and wigwams and takeaways and stuff?"

"I mean spears, blow pipes … And for your information these people have sex several times a week."

"Here!" cried Doris. "Watch it, Professor, there's children here."

Darren and Kelly sniggered.

"What?" said Joe.

"Who? Where?" Frank dropped his Blackberry into his strawberries.

"And they've plentiful beer, practically for free."

"Get me a ticket to the Amazon, Prof," said Joe.

"Make that a double, Igor" said Uncle Frank.

"These fellows have a deal with nature. Or rather, a deal with the spirits which reside in different animals and plants in the forest," said the Professor.

"Spirits?" asked Darren. "They believe in spirits?"

"Yes. They believe that there are spirits which guide them, which oversee them … like Gods. And the spirits are living in different creatures."

"Right. Like Winston would be the spirit of generosity and kindness, wouldn't he," said Uncle Frank.

"Winston might just qualify as the devil, if he weren't eaten by an anaconda," replied the Professor. "The snakes there are very large.

"The people there know what they can take and what they can't. If they start cutting down too many trees or killing too many animals, then the wives withhold favours from the men."

Frank said he had second thoughts about the tickets to Lima.

"How would it be if we lived like them?" asked Doris quietly after a moment.

Igor outlined the deal. He explained that we would only work as much as we needed to get the basics. Joe would be happy just being at home with the family, cooking supper together, doing a bit of gardening together or playing football in the park. A pint or two in the evening. Because of this we wouldn't need a new TV or a second car or a posh kitchen or holidays abroad. We would be just happy pottering about. We would be much less stressed, and so Joe's libi– "

"Yes, quite enough about that, thank you," said Doris.

"And so would yours, my dear."

"Oh my goodness," said Doris, red as one of the beetroots.

"You see, you're all so worried about getting bigger and better cars and kitchens and all that you've quite forgotten yourselves. Those fellows in the jungle aren't like that. They've got no kitchens or cars … they've just got themselves. And they make good use of it."

"But we can't all live in the country like your jungle chums," said Joe. "All those people won't fit in the jungle. There'd soon be nowt left of it. Isn't it better to keep people shut up in cities? At least they're out of harm's way, there."

"Plus once people know what's out there, they never want to go back. You can't squeeze the genie back into the bottle, Igor. You know that," said Frank.

The Professor thought about this. "Well, we'll rediscover what the jungle chums have and transport it back to the cities: the deep, mystical and spiritual bond with nature at the heart of their living. If we rediscover that bond," said Igor, "then we'll find contentment in quietness and simple things … and that will lead to dramatic falls in emissions."

"Bond with nature? What kind of bond is that then, Igor? James Bond?" mocked Frank.

"Sounds more like Tarzan to me," said Joe.

"Of course you don't know it. In most people it's crushed at birth. We can only glimpse it occasionally. It's the joy of a child playing with animals, the deep sense of wellbeing in a sun-dappled forest glade." Igor paused. "It's the scent of mushrooms in the forest, the sound of a babbling stream."

"You get that in most houses in England," said Frank. "It's called damp and bad plumbing."

But he couldn't stem the flow of Igor's poetry. "Have you ever sensed the elemental attachment we have to fire, the thrill of a bonfire on November 5th, the scent of fireworks, or having a barbeque in the garden, the whiff of smoke from the coals? Have you never sensed joy at the smell of cut grass … or the sea wind?"

Joe knew what Igor meant. "When you're out fishing and it's quiet and no-one's around. It's lovely. All your problems go away."

"Exactly," said Igor. "It's rare in the city – but perhaps while gardening a robin hops onto your spade. Or you're walking along the canal and see a kingfisher."

Doris interrupted. "Do you think if a child is brought up close to nature and spends time with animals and trees, he ends up happier?"

"Of course!" said Igor. "If he has the bond, then he's a stronger, more balanced person … and … and he's less likely to need all the things which result in lots of emissions."

"You're talking crap, Igor. Bonds, spiritualism, mysticism. It doesn't exist. It's a dream-world. If people were all lovey-dovey we'd never get anywhere, would we?" said Frank. "There'd be no progress. There'd be no innovation. We'd be sat back happy and not do anything."

"But, Frank," countered the Professor. "We only want progress because we're anxious and discontented. We wouldn't need so much progress. A little, I grant. Moderate, gentle progress. But not progress at all costs."

Frank snorted again. "I've never heard so much nonsense, what with bonds and mysticism."

Igor sighed. But there was a twinkle in his eye when a few moments later he said: "I was just thinking about tomorrow… It's a pity there's a match on. I was rather hoping that we could discuss-"

"Tomorrow?" exclaimed Joe. "It's the Blackburn game! We

can't miss the Blackburn game!"

"But couldn't we just watch the highlights later on TV?" said Igor.

"Are you serious?" Joe cried. "After all you've learnt about football, all the matches you've been to? You can't compare sitting in the lounge in front of the TV with being in the James Hargreaves stand at Turf Moor. It's the atmosphere! Being at the ground ... the lads on the turf in front of your eyes ... your heroes a few yards away. Like you could almost touch them! The roar of the crowd, the chanting ... The whole thing ... It's magical!"

"Magical?"

"Yes, magical!"

"I see," smiled Igor.

"Besides," said Frank. "It's the duty of any fan to go to every game..."

"Duty?" The Professor raised his eyebrows.

"Yeah, that's right. It's their bloody duty," said Frank.

"An obligation? A sense of being tied to the club?"

"I guess so ..." said Joe.

"Obligation," mused the Professor. "... being tied to the club. Sounds to me like something called a bond. First you tell us being at the ground is magical. Then we've your bond with Burnley... I thought you said bonds and mysticism don't exist, Frank? Perhaps I misunderstood you."

"All right, all right, but what's the government got to do with all this?" asked Frank. "It's a free country – whether you worship Brian Jensen or some beast from the jungle, it's nothing to do with the government. Wasn't that why Henry VIII killed all his birds? The government can't force this kind of thing. And if they tried, it wouldn't work. It would all go wrong." Frank snapped open another can.

The Professor didn't reply immediately. He was thinking. "Yes, there may be limits to what government can do. There may be things which policies just can't achieve. Which rather leaves us in the hands of the people..."

"But surely they can do something, Professor?" said Doris. "If all the things we want to do are the opposite of the things we should do to make the world safer for the children, surely the government has to do something. It can't just go on saying we have to borrow and spend on stuff, and at the same time tell us to cut our emissions, it doesn't stack up."

"But Frank's right. It might be beyond what the government can do. They can stop pushing us in the wrong direction but … but, leaders. Now that's a different thing. Great leaders could help us rebuild the bond with nature and recreate the mysticism we need to sustain life and living," said Igor.

Joe turned to him. "So you want them to make us all nicer to each other and to nature, and not so anxious all the time about having more and more and bigger and bigger things. Right? Then life would be a lot easier. And there wouldn't be so many emissions."

"It won't work," said Frank. "You'll never get people to go back to things like gardening and dancing. That's not where the modern world's going. I'd love to see how you're going to do it."

53. Making the bond

How we might go about fixing a deal with nature.

While we watched Strictly, Frank continued to grumble about mysticism and bonds with nature. Later in the evening, after not winning the lottery, he muttered into his whisky that he'd never heard such bloody waffle in his life. He was 47 and what was he doing here listening to an old man from South America rattle on about trees and frogs. "Anyway," he said, "how are you going to do it practically?"

The Professor leaned forward and put his hand on Frank's arm. "Two pieces of legislation. It starts with Do-As-Be-Done-By legislation. You know, an eye for an eye."

Frank looked up sharply. "Eye for an eye?"

"Well, we have to get people to realise we can't just blindly and thoughtlessly trample on the systems we depend on. Nothing could be more stupid."

"Your policies could be more stupid," said Frank.

"All I'm saying is that natural things should have a constitutional right of redress. If you chop down a tree, you should lose a leg. If you cement over meadow-land, it stands to reason you should suffer similar consequences to the meadow. If you are responsible for an oil spill, then an oil-caked guillemot would guide us on justice. If you burn coal…"

"No," said Joe gravely. "It won't work that. People won't take it."

The Professor looked miffed.

Frank laughed. "It's ridiculous. It's pathetic."

"But it's fair, isn't it?" said the Professor.

"Fair? Fair to who? Bloody earwigs? You're barking, Igor. Totally fucking barking."

The Professor poured himself another glass. "Then there are the Undo-What-You've-Done laws."

"You what?" barked Frank.

"Well, what about someone in their forties? I mean, their whole life has been spent plundering the planet. It's obvious that they need to make up for that," reasoned Igor.

"Stands to reason," nodded Doris with a smile.

"So anyone over a certain wealth level would need to invest heavily in forests and bogland and carbon sequestration. It's just an alternative to property tax. Then there's …" And so it went on, a litany of measures to restore our bond with nature. The bulk of them rough justice.

When Igor had finished his long list, Frank shook his head pitifully. "Just one little thing you've overlooked. This nonsense'll put millions of people out of work."

The Professor disagreed. "No more than technology ever did. No more than the loom did to weavers or the steam engine did to horsemen or the typewriter to scribes and the computer to typists and the internet to bookshop employees. Unemployment's never a reason you have accepted for getting in the way of progress."

"But Professor, this really is a bit bonkers," said Joe kindly.

"As it's the only way by which our society can be sustained, you'd better start thinking of it as a fait accompli," replied the Professor.

"What's a fait accompli?"

"It's Latin, Joe. He's showing off," said Frank.

"And who'd make sure we didn't break all these laws?" asked Joe. "I don't see how it would work."

"Who enforces existing laws?"

"The thin green line," said Darren.

"It won't be so difficult. Gradually people will get used to it. It will become second nature. The way we live today will become an ancient saga, a bad dream, part of the lessons of living woven into our culture. When we've restored the bond with nature, it'll just be part of you, like your heart or your liver. Ah-"

He looked towards the TV. Casualty had finished. It was time for Match of the Day.

When it was over, Joe turned to the Professor who was helping himself to another tot of whisky. "Got to do something about this cold," the Professor was saying.

Joe asked: "You weren't serious about all the nature stuff? Chopping people's legs off and stuff?"

"Course he was," grunted Frank. "He's off his effing head."

The Professor looked down at his whisky sheepishly. "Well, if you think that the government should do something about it – not just stand there and wait for Greenpeace and their friends to take the lead."

He was quiet for a moment. Then Doris spoke. "Perhaps there's another way you can make your bond with nature, Professor. Look at Joe and Frank. They were born with Claret blood. In the first few years all they got was Burnley football club. And that bond has stuck with them all their lives."

"Hey, that's right," said Joe. "Forever. It's a watermark in our souls."

"It's a bloody fire brand," said Frank proudly.

"You see," said Doris. "What happens to a child in the first few years of its life – that sticks with them for ever. So, in the first few years, give them nature. Just flood them with nature."

"Put shrubs in maternity wards," said Joe. "No, you'll need live animals there, too," he added.

Frank said that they'd shit on the floor of the hospital.

"Someone'll clean it up. There's your unemployment sorted," replied Igor.

"School visits to the countryside. To farms. To zoos," said Doris.

"Nature reserves. Safari parks," added Joe.

"Parks. Beaches. The seaside. Forests. Mountains," said Doris.

"Rivers, woodland, and moors," said Joe. "Even get them to National Trust gardens. And doing a half day on the allotments."

"Oh yes," continued Doris. "Definitely allotments and garden centres, butterfly centres, marshes, wasteland."

"Send them to the bloody jungles," smirked Frank. "Or ship the buggers off to the desert."

"Wonderful," said the Professor enthusiastically. "Then there's naming, of course."

"Naming?"

"It's obvious that if something has a name, you feel closer to it and treat it better. We don't worry about Africans because we don't know their names. It's the same with nature."

Doris giggled. "You mean-"

"I do. We'll have naming days … when the children are taken out of their schools on trips to name trees. Once trees have a personality, people won't want to cut them down."

Joe shook his head. "This is nutty."

"Not at all. It's a great opportunity for economic growth. Imagine you could only receive state benefits if you visit natural sites regularly. Each child would have a record of natural exposures like its medical records. Imagine the business opportunities around that, Frank. You should be rubbing your hands."

"Bloody bonkers," muttered Uncle Frank. "Sheer bureaucracy.

A massive social experiment."

"Exactly that. So we'd better start as soon as possible. Space needs to be made in the school curriculum for it. It's not an intellectual topic. It's purely emotional and spiritual. Superficial subjects such as information technology and media will have to make way. Children must emotionally feel the spiritual bond with nature and also understand how their actions affect nature and how we depend on it. This must become such instinctive knowledge like when you turn a tap on water comes out."

"You'll have a bloody generation of veggie freaks. We'd be taken over by the Chinese in a trice."

"Did I say we would disband our armed forces?" said the Professor with surprise. "Don't forget that our friends in the jungle have poisoned spears. Death ensues within seconds. We'd surely keep our Trident missiles."

"All right, a bloody generation of veggie freaks armed to the teeth with Tridents. That's even more scary," said Frank. "We'd be a society of lunatics."

"Lunatics with calmness of spirit, balance of mind and self-confidence. Not like the millions of distressed, violent and frustrated urban men. These children would have a far greater chance of contentment than us."

"But what about adults?" asked Doris. We all thought for a while. Then Joe said: "Once I heard that when someone recovers from cancer or comes close to death, it's like there's a reset button. They start their life again with a fresh view. They know what's important again."

Igor clapped his hands. "Exactly! We need to simulate that reset button in the mind of urban man. A country-wide programme of mental detoxification to reset our world-views. With quiet … solitude … proximity to nature and animals…"

"Zoo holiday," said Joe. "Sounds good."

"Right. In bloody Scarborough," said Uncle Frank. "Sounds bloody brilliant.

"I like this," said Doris. "All this would make us less fussed with shopping and whizzing about. More time for you boys to watch football. Frank, what's your problem?"

"I'll tell you my problem," said Frank. "It's bananas. Plus we can't wait twenty years anyway. I thought you said we need action now not wait another generation."

"Twenty years? Why twenty years?" asked Igor back. "We are only talking about the first few years of a child's life. By the age of five or six it would be done; the whole process would only take a few years. And in any case, it's quicker to put people through school than to build nuclear power stations. Don't you believe that schooling works? It must work, otherwise it wouldn't be compulsory, would it now?"

"You know what I mean" retorted Frank. "If you want change, you need to focus on the here and now. Just like us. We've a big match tomorrow, and if we don't win it we're right up Shit Creek."

54. Blackburn

The players were out warming up for the Blackburn game. Joe couldn't make out Mears among them. Frank was talking gloomily to the lad next to them. The latest news was that Mears and McCann had called in with flu. It was all we needed. Frank put his face in his hands. He emerged, eyes red with worry. Things had looked a bit brighter in January and February – we'd picked up nine points, but now we sensed a dull, sickly feeling, the anticipation of failure.

Joe blamed it on our success on the Continent. "It's Europe. It's wearing us out. We're just trying to do too much. It's no use being heroes in Paris if we end up in the Championship next season. The lads are exhausted. They're dropping like flies."

"We're just not up to it," said Frank. He had a very bad feeling.

55. Therapy

A reminder that this is about policy, i.e. what governments might do,
not about what we should all do as individuals.

Frank's fears were not unfounded. The faces in the crowd that
shuffled out of Turf Moor at five o'clock were long and glum.
The two-nil home defeat to Blackburn, our neighbours and rivals,
meant we were stuck on 24 points. Meanwhile Fulham picked up
a rare away win – at Wigan – and shunted us down to 19th place,
a whisker above Hull.

Joe needed a stiff drink. Frank drove us over to his place in his
new BMW X7. We sat at his new retro-shabby-chic bar in the
living room, while he poured out the whiskies. "Top of the range
X7," he said mournfully. "Designer bar. Top of the range
whisky… But it's just an escape really."

Joe explained. "There are times you just have to forget. You
have to escape. However fierce your passion is … it becomes
unbearable. Three losses in a row …"

"It's the only way," said Frank. "Whenever we're on a losing
streak, I just start spending. It's retail therapy."

I could see that the Professor was thinking wistfully of the
immense empty spaces of Patagonia. Its jagged mountain ranges
and ice cold waters. The wilderness.

"You all right, Igor?" I asked. "You look like you're miles away."

"Yes … yes, thank you. I was just thinking about another kind
of therapy. Different from Frank's.

"If the deal with nature works," he mused. "We will naturally
develop a low-carbon culture."

"Culture," snorted Frank. "Culture? You can forget culture,
Professor."

"I warned you," said Joe.

"Not that kind of culture," said Igor. "Low carbon culture
means we won't need therapy. And we'll use our time and money
for other things."

Joe wandered over to the sofa and collapsed onto it, lying back with his eyes closed. "Right, in a low-carbon culture Frank just won't want an X7."

"Exactly. The new X7 becomes triple glazing."

"Triple glazing?" spluttered Frank, caught with his mouth full of whisky. "Instead of an X7? Fifty grand on frigging windows? You're off your bleeding rocker."

"But you'd be just as proud of the triple glazing as you are of the X7," explained the Professor hastily. "It would be absolutely top of the range triple glazing … top of the range … the very best quality money can buy … it would be sparklingly clear to anyone passing by … You could even have a logo on it."

Frank wiped down his jacket with a bar towel. "Are you taking the Mick?" he asked raising a threatening eyebrow. His moustache twitched.

"No, no. Quite serious. Then there's the new flat screen TV. In low carbon culture, if you wanted to decorate the walls, you'd collect paintings. Or even do your own murals."

Frank grunted something else about culture.

"What about Doris's new Honda?" asked Joe. "Where would that go in low-carbon culture."

"It's a ton of emissions just to manufacture it. If you wanted to spend ten thousand pounds on your wife … you'd get her a diamond ring and a bus pass."

Frank knocked back his whisky. "And you bloody think that people are going to do all this just for the bloody penguins. No way Jose. You'll get a few boy scouts and like. But not the masses. They don't give a shit."

"That's exactly the point. I don't think people will willingly spend twenty thousand pounds on insulation and a new heating system instead of a new car. No, this is not about do-gooders and greenies. It's not about boy scouts."

"So what are you saying?" said Frank challenging.

"I'm saying that I'm not hoping that people will suddenly start to be good by themselves. That's why I am interested in what the politicians are going to do about it. What they're going to do to make it happen."

The Professor began to cough again.

"Some water?"

"The boys in the team …" he said once refreshed. "However hard they fight, however much they train … they can't do it by themselves. The gaffer has to get it right. Has to lead and inspire and bully and give them the hair-drier – whatever it takes. The directors of the club have to provide the funds and create a stable environment for the team to thrive in. In the same way … to make rapid, wholesale change in the way we live … policy makers have to be right on top of things, ahead of the game, moving the masses. Look at Leeds … look at Southampton. Where are they now? Wonderful players, but that wasn't not enough."

Joe shuddered. The thought of slipping back into anonymity, just another species which flapped and floundered for a while out of water, and then slithering back helplessly into the mud. That wasn't good enough for Burnley.

56. Buying power

Igor introduces the deal with the rich.

The next Thursday night we were at the Bridge. The back of the Sun was plastered with the news that Manchester City were ready to pay £50 million for Chris Eagles.

"It shouldn't be allowed!" protested Joe. "Rich bastards. It's spoiling football, money is. They think they can just turn up here and nick our best players, just because they have money."

"And just before St Petersburg. It'll totally unsettle the team." That was the other problem. Mr Trew had put his foot down about the trip to St Petersburg. Joe's absences for away games and Europe had stretched his patience to the limit. Even to watch the St Petersburg game in the Bridge he'd had to get off work early, but Mr Trew agreed that was the lesser evil.

Meanwhile the Professor was in a huff about his deal with nature. Nobody took it seriously. He told me that probably meant it was all the more important. Look at Gallileo. Or Darwin. They got ridiculed in their day, too. Now they're common sense. Just as it didn't matter if it took people a few hundred years to accept that the earth goes round the sun or that we are evolved from squelchy things. But emissions are different. We didn't have so much time to get it right.

"Why can't the bloody Russians have the same time as us?" Frank muttered. "If they want to be in bloody Europe, they should stick to our clock."

Igor pointed out that they already have half a dozen time zones. It'd be a bit much to expect them to add two more. "Besides," he added. "When you're owned by Gazprom, you can set the clock to whatever you want. Everyone will synchronise accordingly. It's the prerogative of the rich.

"Rich bastards, as you call them, Joe, have a special position in society," continued Igor once Joe had got in the first round. "Not only do they get their pick of the best footballers, but they also have a special role with regard to emissions.

"And this is why the rich are a special case in emissions policy. First, they consume much more than others – they travel more, they have higher powered equipment, they eat more exotic foods, they have deeper baths, bigger fridges and hotter cookers. Why not? They can afford it. It shows us they're different from the others. This creates their status, the quality which sets them aside from the rest of us.

"Like little birds, little goldfinches," Igor continued, "brightly coloured and singing loudly, announcing that they're superior to the sparrow.

"And second, they have special influence. Rich people control newspapers, run companies, own media and advertising businesses, set fashions. They make laws and they lobby for laws. Their collusion isn't always conscious or deliberate, but just happens in the nature of things. They share tastes in food and art ... so they wine and dine together ... and they exchange views ... and influence each other ... and the views settle on the mutually beneficial."

"Scratch, scratch," chuckled Frank. "third, fourth and fifth rules of business."

"The rest of us imitate their lifestyles – what they wear, where they go, how they travel, what they believe, what they do in their spare time, the houses they build, the food they eat, the cars they drive ... They influence the ambitions and behaviour and choices of many other people. This influence is amplified by their friends in the press."

"Of course independent people like Frank here aren't swayed by the rich ... except perhaps for his BMW."

"Or his Dolce and Gabbana watch," said Joe. "And there's the Paul Smith shoes, while you're at it."

"But most people want to be or feel like a rich person. That's where they want to be," continued Igor.

"You think so?" Doris looked up from her Hello.

"Unfortunately the influence which the rich have ... it's very bad for the planet. They want us to fly more because it makes them richer. They want us to drive more because it makes them richer. They want us to live lavish lives ... have bigger things – fridges, cars, plates of food. It all makes them richer. Of course they want to keep the status quo.

"They dupe us into thinking that chasing those dreams will make us happier."

"You sound like a right old socialist revolutionary, Igor. It's that Russian blood of yours."

"Ukrainian, not Russian."

"Same difference. They don't dupe us. It's not a bleeding conspiracy. It's just, like, if you've done well and you've got an X7 on your drive, it's only natural everyone else wants one. It's human nature."

"Yes, yes," said the Professor. "They don't dupe us. We dupe ourselves. You are quite right. Either way, we need their help."

Frank laughed. "Well you're not having any of my money, Igor, I can tell you that. And a lot of other people will say the same. You can't expect the rich to fund all your naturist fantasies. They've made their money with hard work and they're not going to fritter it away."

"We don't need your money. We need a deal."

"Who's going to do a deal with you? We'll just tell you to sod off," said Frank.

"You're forgetting, Frank. The rich are a minority. And they know it. They're physically weak. And when things start breaking down, it's not the rich who'll have the survival skills."

The volume went up on the TV and the cheers from St Petersberg drowned us out.

57. St Petersberg

St Petersberg was distant and cold and, to boot, had a plastic pitch. And a hostile crowd. By the time our troops got there, two more of our boys had picked up the flu. It wasn't looking good.

The mood in the Bridge was tense. There was foreboding. We watched the huge screen as fire crackers crackled in the night sky above the ground. The pink light of flares shot up from the stands, while heavy booted security men eyed the crowd uneasily. Men with short hair and black jackets jabbed the air with menace. We knew that the Russian attack would be determined. They would wheel out all the weaponry paraded in a soviet May Day celebration and turn it towards Jensen's goal. Cruise missiles, rocket launchers, tank batallions. MIGs overhead.

If we could hold out for a nil-nil draw that would do it.

We did hold out until beyond the seventieth minute. But the superiority of the Russian force eventually told. As the Clarets began to flag, Bikey brought down Huszti thirty-five yards out. The free kick looped in, and the highest man in the air was Sergej Kornilenko whose header thudded in off the underside of the bar.

The beer of two hundred men packed in the Bridge froze. At two-two on aggregate the Russians would go through with the away goal which they picked up at Turf Moor a week ago. The pub was still. The action seemed to pass in a dream.

We fought doggedly for the last twenty minutes. Chances came and went, inches were miles, split seconds aeons. Opportunity did knock, but even as the door was poised to let in a chink of light, Lady Luck, in white and blue, slammed it instantly shut. The lads could do no more. Claret ribbons would not fly on the Europa League trophy.

The dream was over. The glorious battles in Bulgaria, Turkey and Italy, Holland and France ... now they were just fragments in our minds, stripped of their meaning by this defeat in Russia.

The Professor, cold to the drama, was philosophical. "Indeed the rich prevailed."

"Not now, Professor," said Joe. "I don't think we can take any tonight. Let's just get back home." The Bridge was silent. Just the sound of beer dripping from the tap, and the rustling of crisp packets unfolding themselves on ashtrays.

"Remember Frank's advice, my friend," replied the Professor. "Focus on the here and now. Europe was a wonderful adventure. We reached our zenith. But the real challenge is more mundane – it's the challenge of survival. Now the boys can focus exclusively on that. Surely that is a good thing."

58. One rule for the rich

Igor sets out how he wants to turn the rich into a force for cutting emissions.

Chopper came with us to the Wigan game. He drove a brand new Porsche Mephisto. It was the shape of a Cayenne but the size of a Hummer.

"Now that," whistled Joe, "is what I call success. You know," he added, turning to the Professor, "I don't really get how he does it. But that's the proof of the pudding."

We stood in the queue at the DW stadium.

"Still worrying about your deal with the rich, Igor?" asked Frank. He was wearing his black leather jacket, just like Chopper's.

"Sticking at it. Except after how you took the deal with nature, I'm concerned you might think it's a bit … a bit impractical. A bit idealistic."

"Surely not," said Joe. He had a go at an earnest face.

"The deal with the rich recognises the special influence of the rich. It imposes on them certain obligations … in return for which taxes would be kept at tolerable levels. It's very simple."

"More obligations on the rich? There you are, hitting the people you need on your side," said Frank. "They're the only ones that get stuff done quickly. Not corporate types, of course, with their big arses to cover and their lawyers and brands and crap. Easy to act hard with someone else's money. No, real businessmen like me and Chopper. If you leave the world to greenies and fluffies, nowt'll ever happen. They'll just drown in their own bullshit."

The Professor sighed. "We need something to redirect their influence. Retail giants control what we buy and how we live. Banks lure us into debt, trapping us into high carbon living. Consumer goods companies make us believe that our lives are empty without their products. Their messages keep alive the myth that to be happy we have to spend. And spending means emissions."

"So what is your little army going to do about that then, Igor?" asked Frank.

By now we'd passed through the turnstiles and made our way up to the stands. We found our seats.

"Stopping their influence," said Igor. "We ban advertising anything but the most efficient products. You can advertise a Toyota Yaris but you can't advertise a BMW X7. An LED TV but not a Plasma."

"Whoa," said Frank. "No advertising … no TV … no Premier League."

"I didn't say no advertising," said the Professor. "I said no advertising of carbon intensive things."

"Who's going to judge that?" asked Joe.

"He's probably got some quango planned for that one," said Frank.

"I assure you, Frank, I dislike regulations as much as you. But if I just ask Porsche nicely to stop selling carbon polluting monsters, they'll politely ignore me."

"Or even rudely ignore you. Remember, they aren't Englishmen," said Joe.

"So what can I do?" asked Igor.

"I don't know, but we don't need any more quangos telling us what to do and costing us 200 billion quid a year."

"You're right. But think of all the areas of our life where the government could leave us alone. It could leave culture to the cultured, education to the teachers, health to the doctors, safety to mothers, banks to bankers, and bankers to the angry public. In fact there's scarcely any role for government except this," said Igor.

"Now, don't curry favour, Igor," warned Frank. "What I mean is you can't just ban stuff. People need to have a positive incentive."

"Then how about this? Certain people carry great weight on their shoulders. Pop-stars, musicians, film-stars, business leaders, artists, politicians … television presenters, and dozens of inane celebrities churned up by the media industry. And most important, footballers. These people are watched and imitated by millions. Millions aspire to their lifestyles."

"These people create and define 'cool'. Everyone else follows them. So you see it's very easy. If everyone copies these people, then you just make these few people set a good example.

"And there aren't so many of them. They're identifiable by appearances on TV or in Hello. It's easy for the tax authorities to ensure compliance."

At the words "tax authorities" and "compliance" Frank pricked up his ears. He looked alarmed. "Compliance with what?"

"The deal. The deal is that they commit to the very lowest carbon lifestyle possible. Anyone with the misfortune to be a celebrity should have a zero carbon house, an electric car – what's $100,000 to a celeb? No, make that a bike. They could only buy renewable power and … take a vow of vegetarianism six days in seven. There's no need for them to fly. They have sufficient wealth to be able to take their time. In return, they aren't burdened by high rates of taxation."

"Totally unfeasible," said Frank. "It'll never work."

"Of course it will," replied Igor. "They'll soon get the hang of it."

"The hang of it? That's that idea, is it?" Chopper roared with laughter. "I heard about your Sharia law stuff with the penguins."

Igor ignored him. "It's about harnessing their influence. The government can't make things cool. But the rich can.

"So to do away with air-conditioning, they get Penelope Cruz to say: 'I like my men hot and sweaty.' Or Jennifer Lopez to scrap

her car and say: 'I prefer to do it on my feet.' Or Pamela
Anderson to display her dual-cycle ground-source heat pump:
'I'm pumping at home 24 x 7.' Or Laetitia Casta to give up flying,
on the grounds that-"

"Down Fido!" said Frank. "That's all getting a bit sexist, isn't
it, Igor?"

"Oh, I'm sorry to have offended you, Frank. Would you prefer
some male role models? Now, let's see. The Archbishop of
Canterbury … perhaps he could say something about insulating
old churches. Would that be better?"

"Don't be daft, Prof. He's not rich," said Joe. "Frank means
football players. Like all the Liverpool team going vegetarian. Or
Frank Lampard and Didier Drogba taking up gardening."

"State of the pitch out there," said Chopper. "They could start
in Wigan."

59. The deal with Joe

Having failed dismally with the first two deals,
Igor comes to the deal with Joe.

We didn't do well against teams beginning with a "W". Wigan
beat us one-nil, the only goal coming from a penalty early on in
the game before we'd really settled. We were nervous and it
showed in our tempers.

A week later, the last Saturday in March, Wolves came to visit.
They'd had a strong season, were looking for a place in Europe,
and on a miserable, muddy, rainy day, they also beat us one-nil.

We'd lost the last five league games and scored only one goal.
Now we'd need at least 12 points from the last six games. It would
take some almighty effort to survive. Something superhuman.

On the Monday evening Igor popped over to Joe's ostensibly
to talk with Joe and Frank about their bets for the weekend. Joe
was practising his peanut trick and Frank was trying to install his

new BlackBerry.

The Prof was even more depressed than Joe and Frank. "Everything I try goes wrong. I thought the solution lay in the deal with nature. But you all just laughed at it. Then I thought we could do a deal with the rich and bring them in line."

"Dream on," said Frank. "The rich are on a roll and there's no way they're going to play ball with you hopping about preaching at them."

"So we've just one shot left," said Igor.

Joe was fond of the Professor and it upset him to see him so low. Usually he was so optimistic. So many times he'd picked Joe up out of the dumps when Burnley had been misfiring.

"What's your last shot, then, Prof?" he asked.

"My last shot? It's the deal with Joe."

"Me?" Joe missed the peanut and it bounced off his chin into his beer. "You want me to save the planet?"

"The government needs to strike a deal with you … and all the rest of us. We've a hell of a lot to do in very little time. Lots of things need to change. If we just hang about waiting for our emotions to make it all happen – whether goodwill or greed – we'll be far too late. So we just have to roll up our sleeves and get things done. And that means the government needs to strike a deal with you."

60. The Englishman's home

We need to spend a trillion quid making our homes hold heat.

That very evening big news broke: a new plan to renovate Turf Moor. Chopper's grandiose scheme had fallen through, and the new idea was a £20 million rebuild of the current stadium with a shopping centre to boot. Frank preened his moustache as he related the details to the rest of us in Joe's lounge.

Joe was right behind it. "It's high time too. If we want to be a world class team, we have to have a world class ground, don't we? Stands to reason."

"If only everyone were so far-sighted as you are, Joe. As it is, very few people invest properly in their own homes," said Igor, edging the conversation towards his favourite topic. "And by that I mean investing in their insulation, heating and ventilation systems."

Joe snatched a glance at the damp patch by the front window.

"I know," said Igor. "We've already discussed why it's not easy. But the politicians don't have many answers either. The only way the economists have thought of so far – to get people to invest in insulation – is to increase energy prices or subsidise insulation. It doesn't seem to work."

Joe nodded. "You already said that electricity prices wouldn't have any effect at all, will they? Just putting the electricity bills up won't make people invest in their houses – "

Frank interrupted. "Igor, I don't want to be a wet fanny, but, have you any idea what is needed to turn these buildings round?" Frank stood up and paced across the lounge. He pointed to the floor. "First we'll need to take the floor up, dig right down and put in a good 20cm of insulation underground. That's your carpets and your floor boards out. You can put them back after." He wandered across to the front wall, knocking on it with his knuckles. "Then you'll need 15cm of insulation on the inside of all the external walls. So most rooms will need re-plastering and repainting and a bit of rewiring."

"But won't internal insulation make the rooms smaller?" asked Joe. "People won't like that."

"Tough titty," said Frank. "They don't want external cladding either because it spoils their neo-Georgian fantasy. You can't have it both ways."

He tapped the window. "Now, a lot of people have sash windows here." He paused and bit his thumb. "You see your middle class ladies like sash windows but they're total crap. If you want triple glazing – at least on this north side, then you'd best rip 'em out and put in your aluminium and plastic frames. Secondary glazing – not good enough." He shook his head knowingly.

Frank put his hands on his hips and looked up to the roof. "The easy bit's the roof insulation and your cavity wall. And your external doors. They need replacing if you want to keep the heat in. Now, your heating system," he continued. "If you're using solar or heat pumps, then radiators are no use. So you'll have to scrap them and put in under-floor. Best do that before you put your fitted carpets back. Now, you'll have to completely redo the piping. You'll need 400 litres of water storage so you'll have to find space for that. Might need to strengthen the loft or give up the bathroom cupboard. In fact, you may as well redo the whole electrics at this point. 'Cause with all that insulation you're putting in, you've an air-tight building. You stop your heat loss but you'll have a damp problem. So you have you'll need a proper ventilation system with heat recovery. Don't want any accidents, do we? Your LED lights doesn't exactly go with the plastic Louis XVI candelabria, so you'll have to put that up in the attic and get spots. So we'll have to sort out the ceilings. Then you need your low energy fridge, washing machine, dishwasher. Water? Well you'll be wanting to use your rain-water for washing and stuff. So we'll have to fit a rain-water harvesting system – "

"Rain-water harvesting?" said Joe in astonishment.

"Not strictly necessary," said Frank. "But if you want to go the whole hog."

The Professor was scribbling in his book, trying to add things up. He asked Frank: "What do you think this will cost in the average house?"

"Knowing how crap the average house is in England … you're talking 40 grand a go. Maybe I could do it for 38, seeing as you're my mate."

"So where's every single family in England going to get 38 grand from?" asked Joe.

"Er, I think you mean 40, Joe. They're not all my mates, you know," said Frank. "But it's not the money. They can print the money can't they? They spent a trillion of our money on a handful of fucking bankers in London, so they can spend a trillion on this, can't they? That's what it'll take, a trillion quid. But it's not the money, it's getting folk to do it.

"Unless you've a brother in the trade, it's impossible to get someone to mend a pipe never mind fitting solar heating."

"This'll need hundreds and thousands of new plumbers and glaziers and electricians, won't it?" said Joe.

"That shouldn't be a problem," said Igor. "We've two and a half million unemployed people in the country. Some are desperate for work. They'll be retrained and'll have work for ten to twenty years. I know your building profession is as neglected as your housing. The industry is underfunded, under-skilled, unreliable, and unprofessional."

"Here!" said Frank. "It's not our fault folk aren't prepared to pay proper money for our services."

"Of course this'll change," Igor hurried on. "When each household is spending thousands of pounds a year on its property, then the building profession will become as professional as banking, law or accountancy."

Frank scoffed. "Just take care there, Igor. There's plenty of crooks in suits and ties as well."

"And don't forget," said Igor, "that each year millions of pounds are spent luring thousands of students to study university degrees in adventure and media, outdoor leadership, equestrian psychology, aromatherapy, contemporary jewellery or cultural management. These poor students merely become organs of the fantasy game of creating more and more illusions for us to chase. Let employers fund the education of such people, and the public purse support apprenticeships for plumbers and glazers and builders. No, Frank, I think we can find the people."

"It's all well and good finding people to do the work," said Joe. "But who in their right mind would do it? It's a year-long job. Who's got time for that?"

"It's not a year," said Igor. "You can do it in two months." He looked uncertainly at Frank.

"If you're good," nodded Frank.

"Yeah, but you'd have to move out, wouldn't you?" continued Joe. "Who's going to pack up all the stuff, and where's it going to be stored? Where's the family going to stay? Who's going to pay the rental?"

"It'll be in the deal," said Igor. "All inclusive."

"What, kennels for the dog as well?"

"Most definitely. The goldfish, too, I'd expect," said Igor.

"And where do you get all them days off from?" continued Joe. "You can't just have workers roaming around the house by themselves. You'll need three or four weeks off extra, at least."

"Everyone will have statutory time off," said Igor.

"Oh yeah," said Frank. "And small businesses are going to pay for that, aren't they?"

"And how do you know you're not getting screwed? Like you and Chopper wouldn't screw anyone," said Joe, avoiding Frank's eye, "but there's a lot of cowboys out there, isn't there?"

"Of course you'll get screwed," said Igor. "It's in the nature of things."

"Right," said Joe. He was quiet for a moment. "So, why would you bother? The money's not the issue; like you said, they can print that. But it's everything else that's the problem."

Igor stood up and had a little walk around, pulling aside the net curtains to take a look out onto the road. "There's no alternative. Housing causes at least 170 million tons of carbon dioxide emissions a year in the UK. The single most important thing to be done – most urgently – is to make our houses hold heat. Either we knock them down and replace them, or we rebuild what we have. There's about 22 million homes, so over twenty years we need a bit over a million homes a year, about 80,000 a month. Either people do this willingly, or they're forced to." He paused while it sank in.

"Yes. These are very big projects. This is the biggest national project ever. Today in England many people often don't even paint the bathroom when they move into a new home. How do we expect them to devote weeks or months of their lives to rebuilding their houses? You're right, money is not enough. They need a motivation."

"Well, you have to make it cool, don't you?" said Joe.

Igor spat out the word. "Cool, cool, cool! Fun and cool! Cool and fun! In England you think everything has to be cool and fun. Do you ever do anything for serious? Do you ever do anything for genuine reasons? Cool won't work. You can't trick everybody. It's a simple fact. Insulation is very, very boring. Inherently dull and boring. Incorrigibly boring. Irreparably boring. You can paint it pink, put pictures of naked ladies on it, make it sing, turn it into dancing insulation … but it remains boring. Even if Megan Fox is the plasterer, you won't have it fitted. That is the quintessential character of insulation. You can't make it cool. Your politicians would struggle and sweat to make it cool, like a dandruff-flecked comedian at a working men's club trying to squeeze a laugh in the early hours. But they'll never get someone to prefer to spend £10,000 on their heating system rather than £10,000 on a retro-shabby-chic kitchen, just because it's cool.

"And anyway, cool means by definition that only a minority adopt it. If everyone does it, then it's not cool. It can't be cool. And this, my friend, is something which everyone has to do, like washing your hands or learning to read. And don't forget that cool fades – we don't want the refurbishment to be an embarrassment like flared trousers. This must be kept fresh for decades. Once you have a mortgage and children, cool is simply tiresome."

"So if you can't get people to do it because it's worth the money, and you can't get people to do it because it's cool, what can you do?" asked Joe.

"I tell you what you do, mate. You fucking tell 'em they have to," said Frank. "I told you already, softly-softly won't work people in Britain. If they don't have a challenge, if they don't have a poker up their arse, they just sit in front of telly, don't they?"

"What," said Joe, alarmed, "you force them with guns and yobbos?"

"Course you do," replied Frank and he smoothed down his moustache. "It's the only way. That's what Igor means with his contract, don't you see? The only way is to say that if you don't have it done by 2020, then your house'll be knocked down. Then the banks would be on your back all the time to get work done, otherwise they'd lose the mortgage, wouldn't they?"

"That's harsh, isn't it, Prof?" Joe looked concerned.

Frank continued. "It's the threat of demolition – that'll get them moving. It hits you in the guts, you see. Once you see folk on TV in the street, suitcases and like around them, their home being knocked down, not knowing where they're spending the night. That's what you need."

The Professor was silent, perhaps remembering something decades ago. "No, no, no," he shook his head. "Surely it doesn't have to come to that."

Frank retorted: "It will do, I'm telling you. People haven't time, haven't the motivation even if the cash is there on a plate."

"But," began Igor. "You can't do that," he said quietly. "You can't shoo people out of their homes. Well… probably not." He scratched his head. "We don't support Burnley because it's cool. It's not cool to support a football team. And we don't support them because it makes money. We support them because … they're the meaning in our lives. They're our purpose. They're our reason why. So rebuilding our homes has to become our purpose."

He carried on. "Look, the average Briton spends four hours a day watching television. This is because he has no purpose in his life and has nothing to believe in. Their house will become their purpose. It's not about cool. It's much deeper. Their house will be their purpose and their fulfilment.

"My friends," Igor now had the idea firmly in his mind. "The most important things are purpose and desire. If you really want something, you can afford it and make it happen. Desire makes us work better, save more, negotiate smarter, push harder, go to bed later, get up earlier. We have to make people passionately desire a low-energy home. And give them no choice but to have one."

"But what's in it for me?"

"For you, Joe?" said the Professor.

"I thought you said it's a contract. A deal between me and the government. So far all I see is that I have to spend all the time and money I ever make on improving my home. What do I get out of it?"

"Survival, my friend. Survival. That is your purpose."

61. Transport miracles

Igor's plan for quitting our addiction to travel.

We were on our way to the Forest game. We were in 19th position and needed a win to have a chance of staying up. With Fulham away to Manchester United and Bolton at Chelsea, it was a chance for us to move out of the relegation zone. However, things weren't going to plan. The trains were terribly delayed.

"Well at least you've convinced me about transport," said Joe, while we stood on Leeds station waiting for the very late train to Nottingham. "Public transport is a total bloody nightmare. I'm sticking to my Golf. It'll take a miracle to get there by half-time."

Frank was angry. "Miracle? It'll take a more than a bloody miracle. And there's no snow, no leaves, and no frigging volcanic ash either. What's going to be their excuse this time?"

"I am afraid that England still has a lot to learn on the matter of transport. Nonetheless, don't completely discount the possibility of a miracle. Remember what statistics tell us."

"Anyway, we've got nowt else to do. You'd better tell me about your plans for transport," said Joe. "Seems like as good time as any."

Igor didn't hesitate. "As with all cases, we must combine the psychological with the physical. Mind and body in one. I have deep and shallow approaches."

"Makes you sound like a bloody submarine," grunted Frank.

"An excellent parallel. Quiet, effective and unseen. Just as policies should be. My deep approaches are about the psychology of people. The shallow ones – tinkering with technology and infrastructure."

"The psychological battle is to break down their obsession with moving around; and if they must travel, then they will feel revulsion at using petrol-based transport and a strong urge to use clean transport. And the shallow approach – we must simply eliminate the internal combustion engine."

"I see," said Joe. "Sounds clear enough."

Igor nodded. "Let's start by reducing the desire to travel. We must take the mystique out of Mustique. We must prevent the peddling of fantasy. Stop pretending to people that foreign places are nice. Promotion of holidays outside Britain shall be matched by images of endless, hot passport queues, invasive immigration checks, being packed like sardines among the locals on busses without air-conditioning, unfinished seaside apartments, exposed electrical wires and ill-fitting plugs, littered beaches, rude waiters, fat Germans, noisy children, whining mosquitoes and unknown creepy-crawlies, oily food, aggressive motorists, stray pubic hairs, the bad breath of sellers of plastic souvenirs, the scent of vomit, the odour of sweat, panic attacks, incompetent doctors, incomprehensible policemen, incontinent pensioners, impenetrable crowds, and the sogginess of cold chips and disappointment. This will keep people at home."

"You're a real bag of fun, aren't you, Professor?" sniffed Doris. "I like my Benidorm, whatever you say."

"Not at all, my dear. Think of the fun we'll have at home! Investment in local sporting facilities, funfairs, cinemas, biospheres, discos, night clubs, bars, synthetic beaches, water parks, cycling centres, climbing walls … this should keep people at home, no? And I mean places supported by public transport, with excellent bus stops and train stations, and very limited, inconvenient parking facilities. The revival of gardening and allotments will limit the time people wish to spend away. Once you have a productive garden, you're tied to it throughout the summer."

"You've just killed off the transport industry," said Frank. "Not so very clever."

"It's just a question of on-shoring the vacation industry," said the Professor. "Standard business practice."

"Well here's one thing you can stuff in your pipe," said Frank. "Sunshine. Physical and psychological fact: humans need sunlight. And another fact: there ain't any in England. People need it, and there's nothing you can do about it."

Igor muttered something about one trip a year to the Med by train being plenty of sunlight for any sane person. If they wanted more – and he quoted Frank – it's tough titty.

"Oh yes? I think you've forgotten something, Igor," said Frank. "There's no votes in "tough titty"."

Joe said that holidays are one thing but what about getting to work. He pulled down his Burnley hat over his head and hugged himself to keep warm. In a fit of bravado following a tiff with Doris about packed lunch, he'd just come in a Clarets shirt which left a good deal of his midriff exposed to the West Yorkshire wind.

"Yes, yes, I'd not forgotten. Getting to work. The lonely, wandering life of Motorway Man. It's very clear. We must change peoples' priorities – between career ambition and domestic ambition. Many work not because they need the money but because they're infected with ambition and the need for status. It takes them further and further from home like … Motorway Man seeking the deal of the century up the M6 is like a Jew seeking the land of milk and honey. These ambitious people wander further and further away from their roots because they seek fulfilment. But it's a delusion, a simple trick of psychology. They'll never find it outside themselves. The government's job is to correct this illusion. This is the deep approach."

"What do you mean 'correct this illusion'?" I asked.

"I mean to help people feel that what they do is less important than where they do it. There should be no shame in moving to a lower paid job which is closer to home. This should be positively encouraged."

"More ridiculous social engineering," said Uncle Frank stamping his feet in the cold. "It's just ridiculous. You can't hold people back when they want to achieve things?"

"Did I say holding back?" said Igor. "Why can't we be happy being local heroes? Why do we need to be in London for a sense of fulfilment? Isn't it better to be the person who makes Brierfield carbon neutral than the person who writes chapter eighteen point six of a policy document in London? But I am digressing."

"You certainly are," I said. "We were talking about getting to work."

"Ah, yes," said the Professor. "A question of technology. The technology of persuasion and belief-making. We need to make carbon-intensive transport less attractive. We should create an instinctive revulsion to the smell of petrol – perhaps through medical intervention. But until the pills are licensed, we will need more subtle approaches. The motor vehicle must cease to become an object of status."

"You what?" said Joe.

Igor rubbed his hands to keep warm. "The motor vehicle must stop being something people desire. After all, it's not a woman with shapely curves, it's a block of metal painted red. Desiring a motor vehicle is something left over from our childhood. It's simply a form of mental immaturity. Stunted intellectual and emotional growth. People who fetish after motor vehicles have not advanced beyond the emotional stage of infants drawing fire engines or imagining that they're commanding a space rocket. It's hard to correct this intellectual deformity. But it's not impossible."

Frank bristled. "Well, you can bugger off because I'm proud of my X7, and I'm keeping it. It's nothing to do with childhood at all. It's the speed, the power. And the pulling power. If you've a car like mine, you can pull chicks. What bird's going to go out with a bloke that drives a Ford Focus?"

Then he realised he really had put his foot in it, and he looked at me sheepishly. "Now, look what you've done," said Doris, and came over and gave me a hug.

"Sorry mate," said Frank. "Bloody hell! I should have bloody gone to Nottingham in my X7 and not listened to you."

"Possibly, my friend," replied Igor. "But my goal is that society will ridicule the owner of a large car and make him a figure of fun; they will spit invective at him, and he'll become despised; his tragedy will cause embarrassed pity. The Ferrari or the Lamborghini will be admired in museums, on sunny days people will admire their magnificent engineering like the horse-drawn carriage of a Queen."

"He means it, doesn't he?" laughed Doris. "He wants to make Frank look like a twat in his Beamer!" Joe sniggered.

"Watch it, Joe," warned Frank. He wagged his finger at his younger brother. "I'd not get rid of my Beamer even if I did look daft in it."

"And," continued Igor, "we must instil into people a sense of joy at using their own force. We must counter the natural tendency to laziness. This must be done through education, encouragement, exhortation, example and the installation of showers in work-places. We must make a hero of the man who beyond the age of 26 finds himself on the bus."

That moment a chap of Indian appearance approached them. He was wearing the smart uniform of a British Rail employee. On his jacket was a Clarets badge.

"Ladies and Gentlemen. I apologise most sincerely for the delay to the Nottingham train. It was caused, not by technical error or unforeseen circumstances or even the wrong leaves … it was caused by a simple human error. One of my esteemed colleagues forgot to switch a signal, and a small delay was caused in Truro. This led to an unfortunate concatenation of events resulting in the lateness of your train. I hope that you will accept this small token of apology."

He turned to the lad next to him who was pushing a trolley piled high with a heap of fresh, home-made, hot toasted sandwiches and bottles of local cider.

As we looked around we saw that other passengers were being

offered similar refreshments.

"I don't frigging believe it!" exclaimed Frank.

"It's true!" spluttered Darren, mouth full of tomato, and melted cheese running down his face.

We piled into the sandwiches and Doris poured out the cider.

"You see," smiled the Professor. "Miracle number one. But let me continue.

"Then there are so many practical things. They must make local working highly attractive – generic office spaces where people can work in their own villages or towns, the prescription of distant working or telecommuting, mandating of teleconferencing equipment, development and diffusion of technologies to replace the absurd conference industry. Remember that digital technologies can be developed and diffused far more quickly than infrastructural ones. Our transport strategies must become strategies of staying put."

"I don't like the sound of all this 'prescription', Prof. It sounds like bad medicine," said Joe.

"You're right, my friend. But the habit of the car is deeply ingrained in our minds and in our infrastructure. We must take a sledgehammer to it. Just making the car more expensive won't have any effect except inflation. And behind any technical, regulatory, or financial measures we need to change the mind and voices of society so that these new measures become trusted, respected, and aspired to."

Frank had had enough of stomping about on the platform. "When's this bloody train coming? I'm fed up of this place. Last time I ever go to Leeds."

"As to our airports, I recommend their continued abandonment to the free market. This will also guarantee that foreigners will stay away; a policy with which I am sure that Frank will concur."

Frank shrugged.

"I propose the privatisation of our border controls. As a result the borders will be guarded by a crowd of low-paid, incompetent, officious and bullying louts. They'll be unable to communicate in English or any other language. It should reduce incoming traffic to a trickle.

"Public transport must become a political priority – the public can no longer be held hostage to the cosy collaboration of corrupt, cowardly regulators and franchise holders. Policy must play the same ruthless game as the service provider and demand the highest service standards.

"The life of the individual driver must become less and less tolerable. Priority needs to be given to pedestrians and cyclists. Car lanes must be tightened and cycle lanes broadened. Urban road charging needs to be mandated. Sanctions for road crime – fines, bans, and jail sentences for speeding, dangerous driving and drink driving – these need to be trebled and quadrupled. Petrol taxes sustained. We must simply destroy the ugly tyranny of the motorist."

At that moment there was an announcement. "Passengers at Platform 13. The delayed 11.05 to Nottingham has been cancelled due to unplanned precipitation."

We all groaned.

"However, a special service has been arranged. The 12.45 special for Nottingham shall leave Platform 17 at 12.45 in four minutes stopping at Nottingham Station only; that is Nottingham Station only."

It was miracle number two.

"Finally, my friends," said Igor, as the train rolled out of Leeds station, "How to make it attractive to use trains, trams, busses, electric cars, pooling schemes, bicycles, and legs. As we've learnt today, the right choice is not always an attractive one.

"Statutory reductions in work time should be granted to individuals giving up their driving licenses. And large payments for people who give up their passports."

"Is that it?" asked Joe. "Nothing controversial there, for once."

"I think so," said Igor. Then he frowned, and took his notebook out of his pocket. He flicked through a few pages, then leafed back. "Ah yes, I forgot one thing. The petrol-powered car. It'll have to be banned by 2020. From then on only electric cars should be permitted on the road. It shouldn't be painful. After all, the car manufacturers like us to change cars regularly; the banks like us to take out vehicle finance; and people enjoy buying new cars … It should all be a very happy change over."

The loudspeaker crackled. "This is a passenger announcement. For all passengers attending the Nottingham-Burnley game, a courtesy bus will be waiting for you in the station forecourt. The start of the match is being delayed until you arrive."

It was miracle number three.

62. Shopping

"Stuff" causes a big chunk of our emissions, and we need to go easy on it.

"Well that's housing and transport seen to," said Uncle Frank. "Any other aspects of our societies you want to dismantle, Igor?" We were taking a Sunday afternoon stroll through Charter Walk, the shopping centre in the heart of Burnley.

"Several, my friend," replied Igor. "But what if we just stay with two: shopping and eating. They're enough to make a big difference."

"Shopping? Can't we go to the shops any more in your low carbon utopia?" said Frank.

"Not a bad idea, actually, Frank. Think what I'd save if I could cut up Doris' credit card," said Joe.

"Of course you can, my friend. Don't sound so alarmed. There are always necessities which you can shop for. Bread, vegetables, basic clothing." He winked at Joe. "And water, well you can that get from the tap. The problem is shopping for novelty or for status, shopping for therapy, shopping because of boredom. This activity causes you to purchase carbon intensive things – intensive in their manufacture and their use. By cutting down on shopping you'll cut down on emissions. Perhaps 100 or 200 million tons of emissions are caused by shopping for things we don't really need. It will have to go. Or the shops must give us emission free shopping.

"What will all the people who work in shops and the consumer industries do, Igor?" asked Uncle Frank.

"You do keep coming out with this, Frank," said the Professor. "Your concern for others is very admirable. But don't worry about them, they'll be happy. They work half a day, the other half day they keep busy at home in their allotments. Look, they even spend time with their children. I think they'll have quite enviable lives. Not as enviable as that of the people living in the jungle, but they'll certainly have more 'quality leisure' than before."

Joe smiled. "Sounds all well, but how do you stop all this shopping, then?"

"We have to dig very deep."

"In your little allotment, no doubt," said Frank.

"Ready when you are," said Joe, true to his commitment. Meanwhile Kelly had stopped in front of Bargain Zone.

"I told you before, we shop because we're anxious and depressed, we're dissatisfied and we love novelty. We like toys, gadgets, new and shiny things. Like magpies."

"Magpies?" said Joe. "Nowt new and shiny about Newcastle."

"Shopping is the way our mind escapes from the ills of living in cities. Living in cities gives us anxiety, depression, addictions and allergies. And to escape this drudgery we make dreams. To escape boredom we create fantasies. Dreams of kitchens and cars and holidays and dresses."

"No city, no football," said Joe. "It's an urban game, Professor."

"I correct you. Football is played to a very high standard in the countryside of Africa and South America. Now," continued the Professor, "remember our friends the Indians of the Amazon forests."

"Oh yes," said Frank. "The ones who –"

"That's enough, Frank" said Doris sternly.

"They don't have walk-in fridges or flat screen TVs. But the important things – sex and beer – are pretty much the same. Except they get much more of it than you do. I believe that's proof that they have a superior society."

"But you can't stop people shopping, Igor. People need shopping – it's like a social ritual, isn't it. You can't just board up the high street."

"A moratorium is all I ask for. Twenty years of investing and not consuming tat. We must refer to the man in the street as the investor, not the consumer. He'll be an investor because he'll invest in his own home and our future. We must divert our spending from shopping to the home. It's very simple."

Doris shook her head. "Don't you ever think that would be a very dull and colourless world, Professor?"

"Dull and colourless? Because the lights go out in the shopping centres? No, no, my dear. Not at all. It doesn't mean the end of music, art, entertainment, film, fun, good food and drink. It means that we stick with the clothes we have for a few more seasons. That we make do with fewer electronics and keep our laptops and phones for longer. That we go easy on kitchen equipment. More important," he said, "it means that we resist novelty. Because we know it fades in the blinking of an eye. All this the government can change very quickly by no longer indulging the consumer society. It just needs to start by measuring the economy sensibly and not valuing success in terms of how much we spend. Today consumerism is an easy win for politicians."

"But…" began Joe. "Isn't that what the economy is built on? Borrowing and shopping?"

We went into Massarella and sat down for a cup of coffee and some caramel flapjack.

Igor picked up the thread again a few minutes later. "Borrowing and shopping," he said. "Yes, that's today's economy. But there's no reason why it should be just like that. It could also be an economy built on home refurbishment, gardening, singing and cycling. The things people decide to spend their spare time and money on – it's what they see on television. It's arbitrary. Some economies have been based on the most diverse practices and beliefs – consider the construction of churches in the middle ages in England, the wearing of codpieces and ruffs in Elizabethan times or the watching of Antiques Road Show in modern times. Now it's up to governments to take the bull by the horns and change our aspirations."

"It's a big one, isn't it? It's like telling Burnley fans that they have to suddenly start supporting Bolton," said Joe.

"It is."

"It'll never happen."

"We've never tried," said Igor. "But perhaps, just perhaps, supporting Bolton would be preferable to there being no football at all."

63. Food

Igor is looking for a way to get Joe to cut down on meat-eating.

Joe was picking at his tea listlessly. He rolled some peas across to the quorn fritter. He ploughed a chip back and forth through the baked beans.

"Come on, love," Doris said to him. "We're trying to make an effort."

"It's not the veggie food, love," said Joe. "It's…"

"But we're still here, with you, love," said Doris. "Isn't that important? And anyway, you always say we shouldn't cry until the final whistle's gone. It's not all lost yet."

"It is practically," sniffed Joe. He looked up from his phone. "Fulham have just scored."

We were having lunch at Joe's before the Hull game. The family, the Professor and me. Frank was meeting a councillor; we'd see him at the game.

While Joe fidgeted with his fritters and followed the Fulham-Newcastle match on his phone, Doris peeled an orange distractedly. Darren didn't have much appetite either. The Professor, meanwhile, having polished off his main course, hoovered up a second portion of chips, knocked back another can of Carling and vigorously mopped up the rest of his beans.

Through a mouthful of buttery toast and beans he tried to tell Joe not to be nervous. "Look how we perked up against Nottingham. Point by point will do it."

Darren wasn't so sure. "Point by point's all right at the beginning of the season. Trouble is we've five matches left. We need 40 points to be safe, and we've 29. 11 points from five matches means we have to win at least three of the games."

"Where there's a will there's a way."

"I don't think you've grasped the urgency of the situation, Professor," pressed Darren. "At least three wins out of five – it's possible. But – "

"I know, I know," interrupted the Professor, "not when you have to play Manchester City, Liverpool and Manchester United."

"And the last game is against Fulham, our main relegation rivals," added Joe.

Doris began to clear the table, scraping the debris onto one of the plates. "They usually eat so well, my lads."

"There's no point all being so gloomy," said the Professor. "Courage. Determination. Specific action. I am sure the Gaffer has a plan. I bet he isn't just picking at his food, moping. He's out there rousing the boys."

Joe shrugged his shoulders and got up to put the kettle on. "Perhaps a cuppa will make me feel better."

He settled in the sofa with Doris and the Professor sat back into his favourite armchair hands around his mug of tea.

"Food," he began. "It's something that the politicians don't like to talk about. Food and agriculture and cooking are responsible for at least 120 million tons of emissions. That's seven kg per head per day – remember the Smarties? In fact some say that it's much more than that – it depends on what you count. Either way, if we want to cut emissions to 3 kg per head per day, food and agriculture need to give.

"It's not that easy. One of the big causes is meat-eating. Wind from farm animals – at both ends. Especially beef and pork. Chicken's aren't so bad."

"Come on, that's just a joke, isn't it?" said Joe.

"Not if you're an adult," said Igor. "Farm animals give off methane which in the long-term is 21 times worse than carbon dioxide. In the short-term it's about 70 times worse. And it's the short-term we're worried about.

"It also comes from their faeces."

"Faces?" said Joe, puzzled.

Darren explained.

Then Igor continued. "To make a 1 kg of beef we need to grow about 7 kg of crops to feed the beef. It's much more efficient just to eat the crops."

"It might be," said Joe, "but it doesn't taste half so good."

"That might be, but let me finish. Raising cattle takes more land than getting the same amount of food directly from crops. This means more forest and natural land gets turned into farm land. Much of the destruction of the rainforest in Brazil has been for cattle or cattle food. Then there's the emissions from making all the fertiliser we need to grow the crops to feed the animals. And the carbon dioxide from the breathing of cattle. And the energy to cool or warm intensively farmed animals during summer or winter. It even takes more energy to process and cook meat."

"Are you suggesting …" began Joe.

"Sorry," said Igor. "There's no avoiding it. We need policies to get people to eat meat as a treat – once a week … at the very most. It could cut our emissions by about 50 million tons. Well, at a pinch."

"No way," said Joe.

Igor look surprised. "But they eat a lot of fish, there. It's different."

"You what?" said Joe.

"In Norway," explained Igor. "They eat fish."

"Not Norway," said Joe, "No way."

"Got you. There's no alternative, I'm afraid. It's easy to do. It's just about breaking a habit. If we gave up meat we would need to cultivate far less land and could return it to woodland." Then he thought for a moment. "I could go further. Look at the size of people in the street. Don't tell me they need all that food. And did you know we throw away a third of the food we buy? And then there's organic food … But that's all another story."

"So," began Doris. "you're saying that if we only ate meat as a treat – "

"Once a week…"

"Then we could cut our emissions by…"

"By about 50 million tons a year in the UK. Including emissions from imported foods. About 2 kg per person per day."

"Easy really," said Joe. "Just put everyone on veggie crap."

"No harder than all the other things we need to do."

"Go on," said Joe looking at his watch. "We've got to get off to the game shortly. It might only be Hull, but if we don't get three points today, we are as officially stuffed as an organic Christmas turkey."

Igor finished off his tea. He put down his cup and opened up the notebook. A newspaper cutting fell out. He leant forward to pick it up. "Yes, here it is," he said, handing the cutting over to Doris.

She took a look. "That's disgusting," she said.

"Here, what's that?" said Joe, grabbing the picture off Doris.

"It's a picture of how pigs are treated in conventional pig farms."

"It's bloody disgusting."

"This is how to cut down on meat eating," smiled Igor. "All meat products shall carry photographs of the animal being handled at the specific farm where it was kept. There'll be live webcams on supermarket chillers. It's no different from any other product information. It should soon get people eating less meat."

"You and your Sharia law," said Doris.

"Harsh, perhaps. But if you're worried about offending the shoppers, think of the pig."

Doris shrugged.

"Couldn't someone make it cool not to eat meat?" asked Kelly. "Wouldn't that be better?"

"Or rather, cool to enjoy meat from time to time. Yes ..."

"Hey, that slapper on TV with the silicone bazookas," suggested Joe. "'I just do sausage once a week'."

Doris wasn't impressed. The Professor attended to underside of his tea cup.

"Sorry, I was just – " mumbled Joe.

"I'm sure," said Doris sternly, "that there are advertising people who can sort this out for us without needing to dig into your filthy little mind, Joe Sugden."

But Joe was distracted. "You know something, Igor. It's disgusting that farm. Why don't they just ban them straight out? How come it's allowed?"

"I really don't know. Of course it should be banned," said Igor. "It's completely uncivilised and intolerable in any society, low carbon or not."

"Mind," said Joe, "talking about cruelty to animals. There are some endangered Tigers I very much hope we can slaughter this afternoon."

64. Endangered species

We didn't manage to slaughter the Tigers. We'd taken the lead with a lovely goal by Nugent – a screamer from 25 yards at the end of the first half. We thought it was job done. But we couldn't settle, and while we had all the possession, our finishing was nervy. Paterson poked one just wide from six yards. A Fox corner was headed just over the bar when it should really have gone in. And once, with just Boaz Myhill to beat, Elliott had shot feebly straight at the American. Then when the game looked as if was going our way after all, Hull struck. Bullard found a bit of space in our half, made a one-two with Hunt and shot low and hard from the edge of the box. Jensen couldn't hold it, and Geovanni was the first man there to tap it in.

We were sick at having thrown away the game right at the end. Burnley was on the brink. A critically endangered species itself. Now we needed nine points from the last four games.

65. The essence

In the evening Joe had a long call with his dad. It was raining again in Beverley and granddad was in a bad mood. The telly in the home was broken and they couldn't get someone to take them to watch in one of the pubs. So he'd had to listen on local radio which was biased towards Hull.

Afterward Joe consoled himself by watching the Premier League highlights. Joe adored highlights. You could see all the weekend's Premier League goals in half an hour. Goal after goal after goal. Flashes of brilliance or muddy goal-mouth scrambles, the goalkeeper's desperate lunge, the net billowing, the celebrations. Just the pure extract of football, like eating truffles. Well, if you like truffles. Not the chocolate ones, but the mushroom ones which pigs sniff.

"You know," Joe turned Igor. "I could watch highlights all day long. I sometimes think that if I was ever ill, like really ill, like had to spend weeks and weeks in bed, I don't think I'd really mind. I'd just like watch all the goals on You Tube, and when I'd finished them, I'd start all over again."

Igor smiled. "And these highlights … it's just goal after goal?"

"Well, pretty much. They might put in a good save or two."

"The essence, eh? Football boiled down to the essence."

"I guess you could say that. Never really thought about it," said Joe.

"That's it!" exclaimed Igor. "The essence! The essence of football is the goal! Scoring the goal, the joy of the goal, the moment of the goal, the irrevocability of the ball crossing the line, the explosion of delight! Everything's about the goal. Yes!

Yes! Suddenly it all makes sense!" He jumped up, whooped with joy, danced over to Doris and kissed her and then hugged Joe.

"Here, Prof, calm down," said Joe, struggling to escape the intimacy.

After a while Igor calmed down. He became thoughtful again. Then he asked us. "But what's our goal in the fight against emissions? What will bring the flush of adrenalin when we succeed? If we don't have one, then we've no chance."

66. The choice

Either we endure a few decades of authoritarianism
and austerity to cut emissions…

The journey from Burnley to Manchester doesn't take long. The bus timetable quotes 60 minutes. It felt like as many days – like a journey to the dentist. City were in the final of the FA Cup and were chasing at the heels of Arsenal and Manchester United in the Premiership.

"Be brave, love," whispered Doris.

Joe didn't have a word to say. Nor did Darren. Doris went back to her Hello, but she was just looking at the pictures and not reading the words. Kelly was listening to her ipod, drowsily. Frank was going to catch up with us at the match – he was seeing a man in a pub somewhere.

Igor and I chatted for a while – I could see that even he was nervous. He knew this was a big one. The odds were against us. The only thing left was…

"Hard work," snapped Igor. "Forget about luck. Forget about hope. Forget about miracles. It's down to hard work. And … ah! I have said it all before."

"Nervous, Prof?" laughed Joe weakly.

"There's very little time left, my friend. Of course I'm nervous. We are at the point where our options are becoming more and more limited. If we're serious about survival, we no longer have the luxury of choice."

"Neither do we," said Joe. "There's only one option: to win."

It started badly. City fielded a small army of Brazilians, a couple of Argentinians, Spaniards, Paraguayans, Togolese, a Bulgarian, and several other exotic players with outrageous ball skills. They had perhaps fifteen or sixteen men on the pitch at any time. They danced through our defence like another species and were one-nil up after three minutes. They made it two in the twelfth minute – Tevez bundling in a Petrov cross at the near post. We sat on our hands and had bleak thoughts and dry mouths.

There was a "but". We had a bit of luck shortly after the break. A Jensen punt, Bridge slipped and Fletcher kept a very cool head. The feeble flame of belief was lit. Our midfield found its spirit; our forwards found their legs. Elliott skipped in from the right, his pass between two defenders put Blake through in the box, and he smashed the ball past Shay Given.

Within the space of a few minutes we'd become equals to the cosmopolitan acrobats and aristocrats of City. That wasn't enough. We needed nine points and this was the chance to take three of them.

Eagles dribbled past four men on the left, meandering his way to the penalty area and in a flash nutmegged Micah Richards, who caught Eagles' ankle as he went passed him. The referee thought for a split second, consulted his linesman – an hour or two passed – and he pointed to the spot.

None of us watched, except Igor, as Alexander stepped up to the ball. Given got a hand to it, but the stars were with us and decreed that it would be our day. The ball spun off Given's glove and rolled into the net.

There was a strange feeling in the Bridge that evening. Disbelief, as though what we'd seen was such a distant memory that its truth could be doubted. Fatigue, from the sheer nervous energy we'd expended. Merriment. And daring to hope.

Igor allowed himself a smile. He was even wearing a claret

and blue tie.

"You were saying, Prof," asked Joe quietly.

"I was saying that we no longer have the luxury of choice. There are things we must do very quickly – rebuild our houses, scrap our cars, revitalise public transport, transform our agriculture, stop eating meat every day, build nuclear power plants, erect wind farms, shut down our coal-fired power plants, stop using cement and steel ... and much more. And all in one or two decades. No more. The courteous cough of market mechanisms, the timid nudge of taxes, a quiet hint about morality, tarting it all up with ersatz coolness... These aren't policies for making things happen here and now. Good manners!" Igor laughed scornfully. "The well-bred Englishman must make way for his bulldog."

"You mean we just have to roll up our sleeves and crack on. Like we did today against City," said Joe.

"Yes, my friend. We don't have the choice to indulge in aesthetics and special interests, to indulge doubt and standards of proof and all the chance for prevarication it brings. There's no time to wait for conflicting commitments to pass. It's too late. We're past the 90th minute. Our sole duty and purpose is now to cut emissions."

"Hear hear," said Frank. Everyone turned to Frank, astonished. "Well, I mean, like," he stammered. "I mean, I still think you're talking bullshit, Igor, but ... if you want to get stuff done, you can't be pleasing everyone. The only way to get people on your side is to give them the truth. Tell them it's going to be a fucking difficult twenty years. That we all have to knuckle down together. Buckle down. Roll up the sleeves, street by street, town by town. No shirking. Like in the war."

The Professor beamed. "Frank, we're in full agreement. When the government tries to please everyone, it only confuses the man in the street. It must regain its authority. It must give a firm, clear and consistent message. It must project a sense of urgency, duty and purpose."

"That's exactly what I mean," said Frank. "No bullshit. No fannying about. No London liberals or silver-tongued Scotsmen telling fibs that it's going to be easy and fun and win-win-win. Just tell the truth. The fucking blunt truth. That's how to get people to listen."

"It's not going to be such a nice world, is it?" said Joe. "I mean 20 tough years, the hard slog to put things right."

"Not a nice world? With communities working together; with families helping each other out? A society with a common purpose, rich and poor alike? It'll be wonderful. And what's twenty years of authoritarianism and austerity compared to no future at all," said Igor. "But, there'll still be plenty of football and beer. What else do you want? I mean, really want?"

Joe didn't really know. He just felt that something was missing. Frank was fired up at the opportunity.

"Isn't there any other way?" asked Joe. "Something not quite so … forceful?"

"There is. But … My friends, I'm very tired now. I'm not quite as young as you boys. Let's continue another day." With that Igor slowly got to his feet − for once he looked all of his age − and made his way to the door.

67. The last alternative

… or we go spiritual and replace our material needs with deeper purpose.

Three games left. Two wins needed. First up was the home game against Liverpool. We needed some magic and it came in the form of serenity. We played with celestial calm and confidence. We put aside hustle and bustle and sweat and brawn. Our passing was precise and timely. We were infused with the spirit of Barcelona. The defenders mopped up trouble before it became a hazard. And two clinical attacks were enough. A goal on the break after soaking up intense Liverpool pressure sent the home crowd into rapture. Then the stands were almost flung apart by the explosion of joy and cacophony when, with ten minutes or so left, Mears took the ball to the bye-line on the right, and his perfectly weighted cross was headed in by McCann, who'd slipped in unmarked into the six yard box.

"Best I've ever seen them play," beamed Joe. "I didn't think it was possible."

Frank ordered champagne at the bar and we tucked into roast chicken to celebrate. Even the Professor joined in. "Meat as a treat," he quipped. "Celebrate chicken!"

"I guess it depends how often you have something to celebrate," laughed Joe. "Here, what was it you were going to tell us? You said it didn't all have to be so rough ... so aus ... you know."

"Refinding our spirit," said Igor as he demolished a roast potato.

"You mean turning to drink?" asked Frank.

"I mean a different way of thinking. Where we aspire to simple, serene living. We would just use enough to be content."

"Oh yeah," said Frank. "God box stuff? Your chums in the jungle?"

"Precisely." Igor smiled. "Now, this may or may not happen by itself. But we've no time to find out. The only alternative to austerity is for the government to fast-track spiritualism."

"You mean like Iran and stuff?" asked Joe.

"Hmm ... I hope more gently."

"But, Professor," began Doris. "You can't crush people's freedom. It'll all go wrong if you do that."

"But the freedom of the last few decades, it's been destructive, indulgent freedom. Freedom without belief. And belief can only come from austerity or spiritualism If we are to cut emissions sufficiently – quickly enough – then it's either austerity or spiritualism, or some combination of the two. This is the Premier League after all. Liberals and democrats – they'd struggle even in the Blue Square League."

68. Happening now

A reminder that we are already into the end-game.

We returned to Manchester for the second time in two weeks but this time we lost. There was no mercy at Old Trafford. Fulham drew at Birmingham. And so after a couple of whiskies at Joe's, Frank was ready for a fight with anyone.

"Now then Igor, you're a very clever man. I even quite like you. To think a Ukrainian would know the names of all the 1914 cup-winning team. But all this claptrap about austerity and Bible-bashing. Do you really expect anyone to give up steak just to save a bloody penguin?"

"Of course I don't," said Igor. "You are living proof of that, Frank. You're a successful business man. A self-made man. You're independent and you depend on no-one and answer to no-one. You're an island. And ... between friends, and excuse my bluntness ... you've the emotional sophistication of a pit-bull."

"No offense, mate. That's me, Igor. Tough, rugged, answer to no-one. I am an island."

"Don't forget that islands will be the first to go when the sea levels rise. As it is, my friend, no man is an island. Whether you like it or not, you're a part of our society. Just as Burnley is part of the Premier League. If the League falls apart ... what would Burnley do? Play against the local firemen? If the League does well, then Burnley prospers. And in the same way, Frank, you prosper when society prospers. Are you telling me that your building business does well during a recession?"

"Course not, dead as a doornail," said Frank.

"So, you're buoyed by the fortunes of society – you're more a floating rig than an island." The Professor paused. Then he added: "When human societies come under pressure or stress they can break down and horrible things happen. Sometimes that pressure is to do with drought, famines, the spread of disease, the aftermath of hurricanes. Sometimes the economy collapses. And sometimes just because a visionary or a thug wants social change to happen at all costs. Then things like the Holocaust, Pol Pot, Stalin, Rwanda or Srebrenica happen. Under pressure, normal rules and morality are abandoned and people do wicked things to each other.

"Climate change is putting the world under such pressure. Things are getting nasty sooner than the scientists thought. In Ireland's potato famine millions died. When rice yields collapse, what will the two billion people who eat rice do? When the sea levels rise in Bangladesh and across South East Asia, when continents become uninhabitable and people move away in their millions, we won't want to put them all up in bed and breakfasts like in Joe's dream. We'll be overrun by them. Already water supplies from the Himalayas and the Andes are being affected – water supplies which provide hundreds of millions of people with life. Wars will happen, for water, fertile land, safe energy sources. Wars against polluters. Nasty wars, too, where power stations get blown up and oil executives get taken out by snipers. "

"But it won't affect me, will it. That's years off. I'll be driving my BMW in Heaven by then."

"Or your Mephisto in Hell," said Igor. "Then there's a future where our crops can't keep up and bands of people scrabble around the birdless scrub like wild dogs." He finished his whisky. "Yes, they'll be scavenging garbage on Juniper Close, gnawing on the bones of their relatives, because everything collapsed and society broke down. No dignity. No culture. Stradivarius violins thrown on the fire to scare away the wolves."

We were silent. Just then the phone rang. Doris went to answer it.

"Nonsense," continued Uncle Frank. "Who needs stradi-what's it called anyway. Survival of the fittest, is what I say. I'll be fine."

Then Doris rushed, in tears streaming down her face. "Joe! Frank! It's granddad. The floods in Beverley! The sheltered home's flooded out, granddad's room's under water. They've managed to get him out but all his stuff's ruined. Oh hell, love, we've got to get over there. We'll have to bring him over." She buried her face in Joe's arms and wept.

"Fuck," said Frank.

69. The limits of reason

You can't persuade people with reason alone; you need to get them in the gut, too.

Joe and Doris spent a couple of days ferrying granddad and his things back from Beverley and settling him up in the spare room. There wasn't much worth saving; his life's collection of Burnley match programmes had gone, washed down to the Humber and flapping in the mud on Spurn Point.

The week swept by and suddenly it was the Friday before the last match of the season. Tomorrow was the relegation decider against Fulham, and after that nothing for months. We sensed the end of something. The end of a long and unusual journey. None of us had much to say.

"I must go there sometime," mused Igor, browsing through an old National Geographic. "I believe it's an outstanding spot for bird watching."

"Where's that, then?" asked Doris.

"Spurn Point," replied Igor. "Perhaps when we take your father-in-law back. It's just down the road from him."

"Well, that's the difference between me and you, isn't it, Igor," said Frank. "If I want to watch birds, I'll fly to Majorca and set up on the beach. If you do, you take the bus to Cleethorpes."

"No, no," said Igor. "Quite wrong. That's the other side of the Humber, it would be no use going to Cleethorpes."

"Whatever," replied Frank. "To tell you the truth I'm fed up with your birds and animals, whether they're in Cleethorpes, Spurn Point or bloody Timbuktu. It's enough looking after granddad all day, I can do without penguins and polar bears just now. And moths and hedgehogs for that matter. I mean, why should I bloody care?"

"Are you listening, Igor?" he repeated. "I said, why should I bloody care?"

"Care?" said Igor taken aback. "Why should you care?"

"You heard what I said, Igor. I'm fucking fed up of it. All your preaching and posing like some bloody saint; all your bloody philosophy and holy Pope claptrap."

"Here, Frank!" said Doris, alarmed at the brewing storm.

"No, let him," snapped Igor. "Let him have his say."

"Let me? Let me?" shouted Frank. "What am I? Some bloody kid?"

"I just meant…" Igor's eyes flashed, his face red.

"Just meant? Just mental, more like. Deal with nature, my arse. Why should I fucking care about nature? You can just fuck off with your effing tropical rain forests and jungle bunnies. Don't you get it? I'm not interested! I don't care!"

"Of course you don't care, Frank. You're a thug, that's why you don't care. A common thug," said Igor angrily, his lips quivering.

"Thug, am I? You don't like thugs, do you?" taunted Frank. "Too big and scary, right? Well I'll tell you something," he ranted. "No-one else cares either. Joe doesn't and Doris doesn't. They've their lives to get on with. You can forget your palm trees in Juniper Close. So stuff that in your pipe."

"Of course they care, Frank. Don't you, Joe? You do, don't you? Of course Joe cares, Frank." Igor was shouting now. "It's you that's got it wrong, you blasted thug."

Frank stood there and laughed, mocking Igor and all he stood for.

Igor's face twitched with fury. The colour drained out of him and he suddenly became a deathly white. I'd never seen anger in someone like that. It was as if something had snapped in him. "Yes, a thug, Frank," he shouted. "I'll show you how to care. You'll see. You see, if someone's a psychopath, you can't explain to them what pity is. Or a sociopath, you can't teach them what love is. If they're mentally ill, you can't teach them who they are."

Suddenly his voice fell to a hoarse whisper. "Frank, my friend. Not to care about penguins or hedgerows or sunsets or your dog or the Bolivian mountain people – that's your mental illness. You've the sick mind of urban man. I can't argue with you with reason. But I'll show you!"

He called to Winston. The pit-bull trotted over to him obediently, wagging his tail expectantly. Igor suddenly grabbed Winston and held him up to Frank. He was speaking rapidly, hoarsely. "You'll learn to care, Frank, my friend, you'll learn." Suddenly Igor started to throttle the dog. His eyes were bloodshot and full of tears, he stared unseeing. Winston struggled and kicked and scratched and made feeble whinneyings but the Professor was too strong. Winston pissed all over the floor. Frank jumped up to protest but just stood there feebly, somehow paralysed. The Professor's face was flecked with blood as Winston thrashed about with his limbs in panic. Then suddenly Igor relaxed his grip and the dog fell to the ground panting and whimpering and jerking, and he crawled away to Frank, smearing his froth across the carpet.

"You see, Frank. Animals suffer just like humans. They mess the carpet like humans when they sense death. They salivate and groan and whimper like human victims of psychopaths. Now, would you like to come to the zoo with me and watch me drown a penguin, or have you learnt how to care?"

Frank was white. His hands were trembling. He had spilt his beer on his trousers. He sat there, a wet patch spreading over his crotch.

The Professor wept. Tears streamed down his face, and he wept uncontrollably with shock and remorse – begging our forgiveness that he'd had to betray an animal's trust to teach Frank what it means to care. We stood, motionless, scarcely understanding. Then he turned: "Good luck tomorrow, boys." and shuffled out of the house.

"Fuck," said Frank.

70. The prize

Our policy approach in summary: leaving it to the very last moment and to chance.

It was the last day. The winner of the Burnley-Fulham game would stay up, the loser would go down. But Fulham would be safe with a draw.

Igor didn't join us for the game. He must have been watching the match by himself, hidden somewhere among the waves of claret and blue which swelled in the stands. Would the game penetrate his emotions? Did it mean anything at all to him after a year of doggedly following his adopted team? Would he share our desolation and our joy?

Then the game was underway and there was no space to think. We shouted ourselves hoarse until our heads ached with our own chanting. The war-cries froze on our lips when Zoltán Gera stepped past Bikey and shot from 20 yards. The ball pranged in off the bar. That was how it stayed until half-time. We began to have that nauseous feeling of inevitability, as if we were already watching the replays and knew that the worst was going to happen.

The second half was a blur wrapped around by the continuous roar from the stands. We shivered with fear, we bubbled with hope, our bodies were tense and wrought with anxiety. "Belief," muttered Joe. "Belief." He was in a trance. Belief. His hands were clenched. He couldn't shout any more. Doris gripped him. Frank, the superman, the island, shook.

The equaliser came in the eighty eighth minute. Fox's free kick from the left was flicked on by Cort and under a challenge from McDonald, Hangeland headed into his own goal.

The boys flooded back into Fulham's half. In the crowd we screamed through the pain in our lungs and throats. A minute later Blake looked up on the edge of the box, a wall of white shirts in front of him, and with scarcely a back-lift he curled in a glorious shot which Schwarzer clutched at but couldn't reach.

One moment Joe was shouting and screaming with the rest of us, clasped to Doris like wild teenagers. A dance of ecstasy! The next moment he slumped into his seat, dead to the world. Doris bent down and held his head close to hers. All around them folk were dancing and waving their arms, wailing, weeping, laughing, sobbing. Sometime in the future, somewhere miles away a whistle blew and the roar around us became a dense throbbing of sound.

Joe's eyes slowly opened. He looked up and smiled weakly. She whispered to him: "We survived."

"Fuck," said Frank.

71. The note book

A few days later the Professor died. He'd passed away in his sleep at the age of 84. I found his notebook on the floor by his bed. It was packed with his scribbles and diagrams collected over the years in the tiniest writing. Towards the end, the notes on climate change were interspersed with diagrams of football tactics and lists of match results, and tiny spider-like sketches of goal-mouth action. On the last page he'd written:

Dear Joe,

It is time to make place for someone else. I did what I could. I have been an optimist all my life, but now I have little hope. Little hope, not none.
Not every Joe has an Igor and not every Igor has a Joe. I have learnt from Burnley that we can survive if we really want it. But it will take a desperate fight.
I am sorry we could not watch the final game together. I was there.
When I saw the winning goal, I sensed the beauty of its flight. I felt the glory in my veins.
I did taste the primeval scent of survival.
Will I get a season ticket in heaven?

Yours truly,

Igor Rowbotham.

P.S. On the small matter of our bet. You will find an envelope in the left-hand drawer of my desk. The boy's done good, as they say.

Epilogue

One day after tea in the late spring, I went for a walk up Brierfield. It was a warm evening, the scent of grass mowings in the air. People were quietly working in their front gardens, clipping the edge of the lawns and snipping their topiary hedges. A few lads were kicking a football about in the drive.

At the top of the hill was a single house set back from the road in a big garden. As I approached, the sound of barking and yapping got louder. I noticed that the front door opened and someone came out. Three or four dogs followed him out of the house. He bent over to scratch one of the dogs behind the ear. He didn't notice me as I watched him quietly from the other side of the road. He pushed his bike along the path, nosed it through the open gate onto the pavement. He turned to lock the gate and waved to the dogs which wagged their tails wildly, pressing their noses at the wire fence. Then he swung onto the bike, smiling to himself, and pedalled slowly down the hill into town.

Above the gate was a sign, freshly painted by hand, letters a bit wobbly. It read: "Uncle Frank's Home for Stray Dogs. All welcome."

Off the terraces

In the following pages you can race through the arguments in the book without the distraction of screaming fans or goal-mouth scrambles.

What this book is (and isn't)

The probability of man-made climate change is so high and the likely outcome so bad that it is worthwhile taking immediate and strong action to cut greenhouse gases.

Therefore I haven't gone into the question of whether there is climate change or whether it is caused by man (although 99% of climate-change specialists are already agreed): the risks are so high we just can't wait for 100% scientific proof. For a simple video on this see:

http://www.youtube.com/watch?v=zORv8wwiadQ

This book is deliberately not exhaustive. It concentrates on emissions in the "developed world" where most of the emissions come from. It does not cover areas of policy like forestry, the developing world, geo-engineering or adaptation to climate change.

Sources of information, supporting data for the numbers used, and references to other books worth reading, are on the book's companion website: www.uit.co.uk/climate-change-for-football-fans

Part 1 – August to December

1. In the pub

Joe and Igor agree to discuss climate change.

2. Getting started

Re-using your plastic bags is not going to save the planet. It's not "every little helps" but "every little helps only a little". To know where to start it is useful to understand the basics.

3. People not machines

To tame the problem of greenhouse gas emissions, you have to get to the heart of it. To get to the heart you need to find real reasons.

Most analysis of emissions starts by looking at power stations and other physical sources of emissions. When we dig deeper we see that power stations are just responding to the instructions of people. Dig deeper still and you see that people give these instructions in order to satisfy a bunch of needs: sociological, physical, psychological and so forth. You can analyse and categorise these needs in loads of ways. Some are right at the core of what it is to be living and a person, others are ephemeral features of life in 2010 and still others are somewhere in between.

If policy-makers are serious about cutting greenhouse gas emissions, they need to look beyond chimneys and see if these different kinds of needs can be satisfied in less carbon intensive ways. Or see if the needs are really needs in the first place.

4. Dimensions

Emissions in the UK as reported under the UNFCCC are about 600 million tons of carbon dioxide equivalent per year. If you add in all the emissions from the manufacture of imported goods which we are arguably responsible for, then you get to a figure of about 900 million tons. That's about 42 kg per person per day. And we need to get from there to about 3kg per person per day.

5. Finding the reasons

Once we have the two categories of emissions which we can do something about – ones from our everyday routines and ones from occasional big spends – we can dig into the reasons. That means a dull exercise of looking at all the things which we do every day, the reasons we give for doing those things, the things which we don't do, the reasons for that, the things we would do if we could, and so forth. We build up a picture of all the activities and their underlying motivations.

6. In the mind

You can analyse some of the things we do or some of the attitudes we have in terms of what is going on in our mind. And when you look at the psychology of these things, you can be tempted to think that things don't actually need to be as they are. Here are some examples:

Eating meat is partly a question of habit or custom or taste. These are functions of our culture or our up-bringing. There are plenty of people who get on fine without eating meat, so it's not as if we absolutely need a lot of it.

How hot we heat our houses is also partly a question of habit – just not thinking about things. In the 1970s houses were much colder but we were used to it. A preference to put on the heating instead of wear a warm jumper is something in our minds.

Another group of things driven by our psychology is the need to conform, to show we belong, to flaunt our status, to keep up with the Joneses. Sometimes the need might have an important sociological function, other times it's just a waste of time. Clothing is a good example of where these influences come into play.

Transport is an area where the mind can play tricks. We often get into the car as a reflex; sometimes it would have been just as quick to take a bus or cycle somewhere, but we are seduced by the comfort of the car and the misleading idea that it's always quicker.

Our attitudes to spending are also influenced by psychological factors. What we consider expensive or cheap is not based on what we can afford and certainly not based on economic analysis – it's based on what we want to afford, the ads we've seen, and the things our friends have bought.

7. Routine

We can view our lives as a pattern of daily routines with occasions where we break out of the routine. Both these situations are bad for emissions. In daily routine mode we are largely on autopilot, so we don't think about how to save energy or cut emissions; in break-out mode the last thing we want to bother about is saving the planet: we want to have some fun!

One feature of escape mode, is that we get used to the pleasures and comfort it brings. E.g. We go out and buy a big plasma telly. For a while it's special and you feel good about it. Then gradually it just becomes normal. So then you need something even bigger or more special to get the same sense of breaking out of the routine. It leads to a kind of ratcheting up of our emissions.

8. Digging deeper

Four different ways of looking at much the same thing – the inexorable drive for more emissions:

(1) Social changes have put mums to work, boosting the economy and thereby emissions. At the same time it means that kids don't get so much motherly attention and respond by ending up all the more ambitious and needing to fulfil themselves. Call it old-fashioned and not politically correct, but it's a view.

(2) We have evolved to have deep-seated urges much like other members of the animal kingdom. These powerful forces drive us to behaviour which is responsible for ever greater emissions.

(3) You can also look at what goes on in our minds in terms of the

flows of chemicals in our brain: our behaviour is influenced by cocktails of hormones and stimulants. If we were really smart, perhaps we could get the same chemical effects from doing low-carbon things rather than the high-carbon things we do today.

(4) If there is one thing which pervasively warps our minds it is advertising. It is responsible for creating or at least egging on all the ambition which results in high-carbon living.

9. Mirages

If economic growth is about the pursuit of happiness then it hasn't worked spectacularly. By working on people's psychology we could achieve just as much happiness without the need for economic growth and the accompanying emissions.

10. Why it's hard to cut emissions and
11. More on why it's hard

There are dozens of practical reasons why it's hard to cut emissions. It's just darned difficult to cut our main emissions from holidaying, shopping, commuting, and energy use in the home. There's a myriad of obstacles of different kinds: aside from our psychological hang-ups, the technology's not there, you'd look daft, it seems expensive, you can't quit the addiction, you can't find a decent supplier, there's no showers at work, it's too much hassle, our infrastructure is designed all wrong ... and on and on.

To be green needs a lot of specialised knowledge and dedication. Most people don't have it.

And however much cash or enthusiasm you might have, a lot of things are not under our control. We can't individually change infrastructure or public services; technologies aren't available at individual scale. Some solutions can only be effected by government at different levels.

12. The PSV game

(On the terraces.)

13. Uncle Frank and democracy

There's a standard argument to defend individuals not doing their bit: "what I do won't make any difference anyway". But (as the Professor points out) a committed fan believes it's his duty to go to a match – because the crowd is made up of the sum of individuals attending.

The Professor also challenges the argument that "solving the problem" is up to government and companies. He says that governments and companies only respond to what the man in the street wants. But Frank points out that if you want to get stuff done, giving people freedom of choice is not very effective.

14. The distribution of power

Society is structured in a way which makes it hard to change. There are powerful groups with vested interests in the status quo. Because we live in centralised societies, politicians, the press, and business have exceptional power over us. The three groups depend on each other. It is hard for any individual with such power to break ranks and challenge the status quo, because to break ranks means loss of the very status or wealth which you fought for all those years. So ironically the very people with power to change things are unlikely to rock the boat.

Frank's view is if you want to cut emissions then entrepreneurs and financiers are the people you need to get stuff done. So there's no use pissing them off.

15. Instincts

This small argument is just to say that there's no point relying on people's instincts to do something about the problem. It's a point worth making because many of us like to do away with government meddling; but if you can't rely on people to solve the problem themselves, then meddling becomes necessary. In short, nature did not equip us with instinctive tools to identify and respond to long-term threats. We're stacked with instincts for handling immediate threats, but not for sensing levels of carbon dioxide in the air.

16. Culture

If instincts can't help us, can culture help us instead? Cultures are good at addressing things which our instincts aren't so good at: like how to behave in a society, how to educate and how to farm. But they still aren't good at really long term things which cross many generations. Most of our cultural rules are about how we should behave towards other people. We were never threatened with environmental catastrophe in the past, and so we didn't get round to building into our culture special rules for looking after the environment.

Some aspects of culture might be useful for protecting the environment, such as the ability of cultures to get people to cooperate and be considerate to each other (as in some religions). But the promotion of individualism in a competitive, winner-takes-all society tends to undermine cultural guidance and makes us revert us to instinct.

17. Values

Sometimes we are tempted to say "Why don't we just... ban this, stop that, change this..." One of the things which stops us "just" doing things is that we are constrained by a lot of values which hold together our civilisation: liberty, rights, justice, peace, equality, dignity, compassion and so forth. These actually get in the way of cutting emissions. You can't force people to do things, you can't withdraw people's rights, you can't limit people's freedom except in very specific ways, and you can't resort to violence. In rare cases politicians are allowed to override these values – in emergencies or during wars.

18. Disasters

Sometimes people say that disasters will prompt us into action. Several conditions are needed for disasters to work. First, they need to be close to home for us to take notice. Second, we need to know clearly the cause of the disaster so that we blame it on the right thing. Third, disasters needed to be repeated, because our memories fade. Fourth, they shouldn't be repeated regularly, otherwise we will just adapt to them. By the time disasters like that come along, it will be too late.

19. Other animals

The thesis of the General is that the only way to get people to start treating the environment properly is to get them to value properly other species of animal with which we share the planet. If we respect and can identify with the creatures which suffer as a result of our activities, we think more about what we do. To engender that sense of respect in people you would have to start instilling it into very young children during their socialisation.

20. Vested interests

There are many social groups which have to be won over in order to get change. This includes variously (and not exhaustively) the energy industry, people who put being clever before being right, people who prize economic growth above all else, people who don't care about nature, people who can't let go of ambition, people who think we can solve all the problems with technology, and people who just don't care about anything. (That's most of society.)

21. The thin blue line

In some cases we accept that economic markets are not the only way to guide society. The strong arm of the law is not a market force; in some cases the forces of law and order are an attractive option.

22. A hard sell

People frequently excuse inaction on some matter of climate policy by saying that it is a hard sell. This section briefly ridicules this defence. It exposes the distinction between the rare politicians who are leaders and the majority who are followers.

23. A brush with the carabinieri

Back to law and order, the Professor reminds us that in our society the guarantors of the government are the armed forces. The government needs to know that the armed forces are behind them. This means if you want to have tough policies on climate change you have to persuade the senior people in the armed forces. Already senior figures in the US army have called for action on emissions. Their views will influence governments which are lagging behind.

24. When in Rome

The EU might have plans to cut emissions, but you need to get the USA and China on board, and half a dozen other countries with large populations. Most of them have no genuine interest or ability to address seriously our predicament.

25. When in Bonn

There is an international effort to cut greenhouse gas emissions, led by the UNFCCC, an arm of the United Nations. Unfortunately it is unlikely to achieve anything useful whatsoever.

26. The time it takes

Ironically we have known about the greenhouse effect for over 180 years. It takes this order of time for ideas to become accepted, turn mainstream and for anyone to do anything about it. King James II figured that smoking was dangerous in 1604 and only now, 400 years later, is there systematic legislation to cut down on smoking. Don't hold your breath.

Modern society is wedded to the method of scientists. But the standard of proof set by science is not practical for political and social purposes.

27. Overwhelming complexity

Greenhouse gas emissions are part of a fiendishly complex set of predicaments. It's not right to think of it as a problem with a solution. There is no such thing as a solution. There are just things we can do to cut emissions (and potentially adapt to the effects of climate change). And in doing the right thing, we can actually make life a little bit better for ourselves and other creatures.

28. Real progress

(On the terraces.)

Part 2 – December to February

29. Christmas crisis

A summary of Part 2 of the book: our efforts to stem emissions are feeble and patchy.

30. The wrong targets

We are trying to set targets to reduce emissions by such and such a percentage by such and such a year. E.g. 50% by 2050. It gives us a 50-50 chance of avoiding catastrophic climate change. Numbers like this are easy because they are meaningless. We should imagine what that means in practice.

Long-range targets leave so much wiggle room that they don't guarantee sufficient overall cuts in emissions. And the way they are set is arguably unfair. Finally, the idea that emissions targets are useful for guiding investors doesn't wash if you are even slightly cynical about why people make investments.

31. Marbles

The way the greenhouse effect works, very crudely, is the more greenhouse gases you put in the sky, the hotter it will get. It doesn't really matter when you emit the gas, the key thing is the total amount you send up. Our targets should be based on this, but today they are not. In short we can afford to emit about another 750 billion tons of CO_2 and then we are stuffed. Full stop.

32. Efficiency and redundancy

When policy-makers try and figure out laws for cutting emissions, they like laws which are "economically efficient"; that is they cost as little as possible. You wonder why it is clever to skimp on addressing what is the biggest predicament we are facing. This exposes how stupid most economists are.

Redundancy means having back-up plans in place in case your first plan doesn't work. That would be sensible since we are talking about the survival of humanity as we know it.

33. Resolution

Emissions come from every nook and cranny of society. If you want to cut them by 80% then you have to go after practically every source of emissions. There isn't a category of emissions where you can say "Oh, that's not important we can leave that." Put another way: if you focus on the main categories of emissions, like power generation and home heating and transport, and even if you cut them down to zero, you still have to cut emissions from other areas to get 80% or 90% reductions. And if you have any slippage in the main categories, then you have to carpet bomb emissions to achieve 80% or 90% reductions.

34. The wrong comparison

When we say that it's "too expensive" to cut emissions, we get our sense of expensive by comparing the costs of cutting emissions with the "value" of things in our economy today. But by definition at least half of our economy consists of things which we want to get rid of. So why would their current value be the measure against which we evaluate our policies? Once we have cut our emissions by 80% or 90%, our society will be so different and unrecognisable, that lots of the things which we value today (like cement plants) will be gone anyway. If your world is an egg and you want to make an omelette, do you count the cost of breaking the shell? Or do you look at the value of the omelette?

35. The wrong numbers

This harks back to earlier points about focussing on real reasons. The inventories of emissions which countries send to the UN each year list emissions according to their immediate source – power stations, industrial facilities, farmland etc. The Professor argues that to be useful the inventories should focus on the causes of emissions, not their technological sources. The argument is a bit tongue in cheek, but you have to wonder what the scientists and economists can usefully do with the massive lists currently produced.

36. Winging it

The focus of current policies is to replace the technologies in our industrial infrastructure. In three decades they assume we can replace or retrofit something like 30,000 power plants and industrial facilities. Plus 600 million cars and hundreds of millions of houses. Unfortunately, except for cars, these are things where new technology spreads very slowly. It's highly unlikely that any policies could realistically result in wholesale replacement of these technologies in so short a time.

37. A new stadium

To cut emissions through technological change we will have to implement thousands of industrial projects across the world in short order. Yet the practicalities of completing a large industrial project are daunting, the timescale is many years, and the process is fraught with risk and delays. You wonder how feasible it is to base policies to cut emissions on an approach which depends on implementation of such projects on a global scale.

38. The Great White Hope

Politicians are placing faith in a fix called carbon capture and storage. This is a technology for capturing carbon dioxide emissions at the exhaust end of a power station and pumping it into storage areas underground. It is an unproven technology and it will be about 20 years before it is viable. After which we will need to build one unit a day around the world for 20 years. So to dig us out of the hole we are effectively making a 50-50 bet and relying on something which doesn't yet work.

39. Write-offs

Most of the power plants and industrial facilities in the world will have to be shut down or retrofitted in the next 30 years, before the natural end of their useful life. As a result of this several hundred billion dollars worth of industrial assets and associated bank loans will have to be written off. A nice problem for governments which have just completed a several trillion dollar bail-out which they could ill afford.

40. Price of a ticket

Making companies and people pay for their greenhouse gas emissions is at the core of much climate change policy. Putting a price on emissions makes carbon-intensive activities more expensive and makes low-carbon activities relatively cheaper. The theory is that carbon pricing – whether through taxes or a "cap-and-trade" scheme" – will get companies and people to switch away from using fossil fuels. It also gives investors a signal about what to expect in the future.

For the carbon price to be "right" – i.e. to be at the level which will result in cutting emissions down to a target level – the target needs to be right. However, industrial lobbying will always ensure the target is too lax and so the carbon price will always be too low to achieve its aim.

41. When money's no object

Making people pay for greenhouse gas emissions might not actually make them change their behaviour. If you pay your fuel bill by direct debit you probably won't know how big it is. Most people can't understand their bill even when they look at it. And if they wanted to cut down energy consumption they wouldn't have a clue what to do. Generally, our habits are so deeply ingrained that for carbon prices to jerk us out of these routines they would have to be so shockingly high as to be politically unacceptable.

Put another way, putting a price on carbon to change people's behaviour is like increasing death duties in order to cut mortality rates.

42. Pricing and industry

Emissions markets might make industry cut emissions if the markets where those industries operate were "pure" markets. But the energy sector is so politicised and based on cronyism, and hard for new blood to enter, that the sector is unsuitable for a market-based mechanism like trading. Perhaps a few countries in Europe have free markets for energy, but 90% of the power plants in the world aren't in such markets.

Emissions trading might work eventually. But we need to cut emissions in the short-term not in the long-term. For short-term cuts we cannot rely on emissions trading; we will have to look elsewhere.

43. Signals

Does carbon pricing work as a long-term signal to investors? The trouble is that the signals come from politicians. As these are among the least reliable people in our society, the signals themselves are highly unreliable. And if they are unreliable, then no-one will pay much attention to them. Another problem with market-based signals is that in a market prices go up and down. So one moment the carbon price is telling you to invest in renewables; the next it's telling you to go back to coal. This is unlikely to make investors confident about investing in the low-carbon economy.

44. False choice

The main policy of the EU to cut emissions is the EU Emission

Trading Scheme, a "cap-and-trade" scheme. If a company in the scheme is short of allowances to emit carbon dioxide, it has a choice: it can either cut its emissions or it can buy in allowances. This is often presented as a real choice, but in fact it's like the choice between learning German to read Goethe in the original or just picking up a translation: who's going to bother embarking on a many-year project to cut emissions when they can just go into the market and buy allowances?

45. Le prix d'une bière

Some economists like the idea of a global carbon market. They say that this will allow us to cut emissions at the lowest cost to the economy. For them the objective is to cut emissions as cheaply as possible. Unfortunately their objective is wrong. We should be cutting emissions as quickly as possible, not as cheaply as possible. If we want to cut emissions as quickly as possible then we don't want just one price. The price which makes a cement plant cut its emissions is too low for an airline to worry about its emissions. And the price of carbon at which an airline would cut its emissions would put the power sector out of business on day one. So it makes more sense – if you want to cut emissions quickly and you insist on carbon pricing – for each industry to have a carbon price set at a level which stimulates it to cut emissions. Then you get effort throughout the economy to cut emissions. People who imagine that we can cut emissions quickly with a single carbon market and a single price across the economy, are doing just that: imagining.

46. The madness of long-termism

Worrying about 2020 and 2050 is a distraction. We need to abandon long-term policies and adopt ruthlessly short-term policies in the matter of cutting emissions, because we have only two or three years to stabilise emissions or else our emissions budget will quickly be exhausted. We need to focus on 2011 and 2012. Luckily, there are lots of businessmen who are excellent at thinking in the short-term, so we should be OK.

47. Praying

Some economists are dismissive of spiritual things like mysticism and faith. They are misguided. Current policy thinking is caked in huge dollops of faith and mysticism: that we can replace the infrastructure around the world in a few decades; that engineers will get carbon capture and storage to work; that technology will "save" us; that industry and consumers will respond to carbon pricing; that markets will work. They might as well get out and do a rain dance.

Part 3 – March to May

48. A night on Pendle Hill

We are like locusts on a field of wheat. We are stripping the planet bare. Democracy gives us the freedom to eat as much as we want and more, and fear of war means we can't do anything about it. In short, we are stuffed.

49. Joe's dream

If millions of "climate change" migrants start coming here looking for somewhere to live, we might need to build bigger spare rooms.

50. The first years

The way we are is determined in the first few years of our lives. After that, we are less influenced by the things around us. The place to start is educating our children.

51. Vision

Policy-makers like to sit around creating visions for the future. If you need to look busy, it's a good one. Time for visions is up. Focus on the next game.

52. The Burnley Protocol

The Burnley Protocol is some back-up in case we can't replace all the power stations quickly enough; in case carbon capture and storage doesn't work in time; in case trading schemes and taxes don't have the effect we hope. The Protocol contains three deals. A deal with nature. A deal with the rich. And a deal with Joe.

53. A bond with nature and
54. Making the bond

We need to create a deal with nature to live in harmony with her again. Whether you have communism or fascism, liberal democracy or a libertarian paradise, human nature will destroy the planet unless there's a contract with nature. If people are imbued with a deep spiritual respect for nature it does two things. First, they are less likely to slash and burn because they respect it. Second, they are going to be calmer and more sorted people, ready to live simply and without all the shit in our lives which causes emissions.

There are two approaches: a tough one and a softy one. Under the tough one, if you fuck with nature then you get it in the back. Under the softy one, you start kids on nature when they are very young and just flood their minds with animals and plants. It sounds daft but it's just common sense.

55. Blackburn

(On the terraces.)

56. Therapy

A reminder that this isn't about what we should be doing as individuals. It's about policy, i.e. what the government ought to be doing.

57. Buying power

The deal with the rich says that the rich have a very special place in society and in the problem of greenhouse gas emissions because of their power and influence over everyone else.

58. St Petersburg

(On the terraces.)

59. One rule for the rich

Somehow we want to keep the energy and creativity of the entrepreneur but channel the influence and power of the rich in the right direction. The most malign influence of advertising needs to be emasculated. More importantly, the deal with the rich gives them a choice between a green lifestyle or high taxes. Simple really.

60. The deal with Joe

If you don't like the deals with nature or the rich, then it's down to the deal with Joe. Each of us just has to roll up his sleeves and crack on.

61. The Englishman's home

The single most important thing to be done – most urgently – is to make our houses hold heat. This is a one trillion quid deal to encourage or force everyone to invest in their homes and   massively cut fossil fuel use: whole hog insulation, ventilation and renewable heat generation. It's tough titty time, meaning that a market-based approach is too late and that some coercion will be required from government.

62. Transport miracles

Obviously, stop supporting cars and planes, and get people onto their feet and into trains and busses. Stop peddling the fantasy of travel. Invest in making sun-seekers content with their allotments and motorway man satisfied with local work. Pay people to give up their passports and driving licences. Break the obsession with motor cars through ridicule as much as taxes. And if that all fails (which it probably will), make the electric car compulsory in short order and build nuclear plants and solar, hydro and wind farms to charge the batteries.

63. Shopping

A twenty-year moratorium on consumerism would switch our spend from the shopping mall to the DIY and gardening store. Instead of spending on crap we will invest in our homes and a low energy life-style. Concerned about the people in the consumer industry? They need all the free time they can get to learn how to fit windows and grow potatoes.

64. Food

If junking consumerism was easy, the food issue isn't. We have to cut down massively on meat (and probably dairy, too). Meat as a treat, once a week, is the maximum we can afford. It might be enough to expose people to how animals are treated in industrial farming to get them off the meat habit. So on the packaging you put pictures of how food-animals live.

65. Endangered species

(On the terraces.)

66. The essence

In football the goal is an orgasmic moment – that's what football is all about. It embodies the purpose and motivation. Without finding an equivalent in climate policy, there is little hope for change on the scale needed. The search goes on.

67. The choice and
68. The last alternative

Liberal democracy won't help us cut greenhouse gas emissions. We have a choice: either an austere, authoritarian regime where in the next twenty years we jolly well have to get on with it, or aspiring to simple, serene living like the Amishes in the USA or indigenous people in the jungle. Time for politicians to get their thinking caps on.

69. Happening now

And they don't have much time for thinking because scary things are already happening.

70. The limits of reason

Reason and logic and intellect aren't what they are cut out to be. They don't persuade people. In the end what persuades people is visceral things. We shouldn't shirk from this unpleasant fact.

71. The prize

Sorry, that was just the Hollywood version. It didn't really end like that. You didn't think that cutting emissions would be that easy, did you?

But one bit is right: we are into the last minute and we are relying on long-shots from outside the box.

72. The note book and
73. Epilogue

Hollywood continued.

Acknowledgments

Many thanks to Paul Bodnár who suggested I write a book; to Barna Baráth, László Pásztor and everyone at Vertis for allowing me the time off; to Jason Greenwood, Stephen Heal, Emily McClave, Bob Mullan and Siobhan Parkinson for their thoughtful reviews; to Flóra Ijjas, Martin O'Connell, Alastair Sames, Dale Southerton and Guillaume Wolf for their help and input; to Andy Skelton for his support and research; to Niall Mansfield, my editor at UIT; and to Zsófi, Imola and Tim for their patience.

Website for this book

Register your book: receive updates, notifications about author appearances, and discounts on new editions.
www.uit.co.uk/register

News: forthcoming titles, events, reviews, interviews, podcasts, etc.
www.uit.co.uk/news

Join our mailing lists: get email newsletters on topics of interest. **www.uit.co.uk/subscribe**

How to order: get details of stockists and online bookstores. If you are a bookstore, find out about our distributors or contact us to discuss your particular requirements.
www.uit.co.uk/order

Send us a book proposal: if you want to write – even if you have just the kernel of an idea at present – we'd love to hear from you. We pride ourselves on supporting our authors and making the process of book-writing as satisfying and as easy as possible.
www.uit.co.uk/for-authors

UIT Cambridge Ltd.
PO Box 145
Cambridge
CB4 1GQ
England
Email: inquiries@uit.co.uk
Phone: **+44 1223 302 041**